Human Caused Global Warming

THE BIGGEST DECEPTION IN HISTORY

*The Why, What, Where, When,
and How It Was Achieved*

DR. TIM BALL

Tellwell Talent
www.tellwell.ca

ISBN
978-1-77302-130-0 (Paperback)
978-1-77302-131-7 (eBook)

Contents

Preface

There is a big problem once you decide to use science for a political agenda, especially if the scientific hypothesis is untested. If it fails when tested going forward, the only options are to admit it doesn't work or make the evidence fit the story. The people who formulated the human caused global warming hypothesis chose the latter and the deception began.

This is a handbook written for the general public providing information that they don't hear from their government or most of the media about how the greatest deception in history occurred over the last many years. It is the story of how and why the global warming deception was achieved. The world has not warmed for over 20 years, yet carbon dioxide (CO_2) levels continue to rise in complete contradiction to what all governments are saying. And that in itself should make people ask a lot more questions than those identified in the book.

This book explains the entire story of how and why global warming, the claim that human activities produce CO_2 that causes temperature increase, as a scientific issue was exploited for a political agenda. It is presented in the form of a criminal or journalistic investigation answering the basic questions, Why, Who, What, Where, When, and How.

Footnotes or references are not included, but all the material is available on the internet. If the reader wants expansion of any section or reference material, it is available in my book *The Deliberate Corruption of Climate Science* shown at the end of this book and on my website at drtimball.com.

Over the years, my challenge to the government version of global warming became increasingly problematic. Neither the government nor the environmentalists could say I wasn't qualified. I am blessed with an ability to teach and explain complex ideas so people can understand. This talent was honed by teaching a Canadian science credit course to arts students for 25 years.

The attacks against me include death threats, large amounts of false information about my qualifications posted on web sites, including Wikipedia, and three law suits, two of which are still pending. Most people cannot believe that such things occur in a supposedly democratic open society. If you want to test the idea, just say to people that you don't accept the idea that humans are causing global warming. You will be amazed at the vehement reaction you get from the majority of people, all of whom know nothing about the science.

We live in the delusion that we know what is going on, but we don't. Author and medical doctor Michael Crichton described the problem in the opening paragraph of a 2003 speech he gave in San Francisco.

> *I have been asked to talk about what I consider the most important challenge facing mankind, and I have a fundamental answer. The greatest challenge facing mankind is the challenge of distinguishing reality from fantasy,*

truth from propaganda. Perceiving the truth has always been a challenge, but in the information age (or as I think of it, the disinformation age) it takes on a special urgency and importance.

In this book I am not asking you to believe what I am telling you; people are tired of being told what to think. I am simply setting out facts you have not been told by the government or environmentalists, so you can ask yourself why. Then, put together what government and environmentalists and the mainstream media tell you, and draw your own conclusions. I often begin a presentation by saying "Morning." Then I point out that I didn't say "Good morning" followed by "Make up your own mind."

Dedication

This book is dedicated to my wife, Marty, and my sons, Dave and Tim Jr. My pursuit of the truth and a willingness to speak out triggered Voltaire's warning. "It is dangerous to be right in matters where established men are wrong." Unfortunately, members of my family were also victims of the attempts to silence me. For that, I am forever grateful for their unending support and the sacrifices they made.

Human Caused Global Warming

THE BIGGEST DECEPTION IN HISTORY

The Real Consensus

Hands-up. Who thinks greenhouse gases have no effect, and
therefore we all need new jobs? Anyone?

Description of the machinations associated with the story of global warming as the "biggest deception in history" is not hyperbole. Some call the claim that humans are causing global warming a hoax, but that supposes a humorous objective. Earlier deceptions did occur but not on a global scale with negative global impacts. A deception is a deliberate act to mislead – the objective is to provide information predetermined to support the claims of impending disaster. The deception is the hypothesis that human production of CO2 is causing global warming. The hypothesis is referred to as Anthropogenic Global Warming (AGW). The agency that carried out the deception was the UN Intergovernmental Panel on Climate Change (IPCC). Very few people read the science reports they produce, and even if they do, few understand. Most scientists assume that what other scientists produce is uncorrupted, but look what German meteorologist and physicist Klaus-Eckert Puls found when he looked:

> Ten years ago I simply parroted what the IPCC told us. One day I started checking the facts and data – first, I started with a sense of doubt, but then I became outraged when I discovered that much of what the IPCC and the media were telling us was sheer nonsense and was not even supported by any scientific facts and measurements. To this day I still feel shame that as a scientist I made presentations of their science without first checking it.

There is nothing funny about the damage and cost to people and economies. Some put the cost in the trillions. The people who produced such a report knew what they were doing. Deception on such a scale requires planning. It needed an objective, a process, a structure to carry out a plan that fooled most of the world. I say "most" because there were a few of us who knew what was going on.

Personal attacks and the marginalizing of those who knew and spoke out is evidence that the global warming charade was a deliberate deception. Increasing exposure of the deception resulted in an intensification of the attacks. Global warming skeptics became climate change deniers, who were threatened with lawsuits, including charges under the Racketeer Influenced and Corrupt Organizations Act (RICO) designed to deal with organizations like the mafia. As Shakespeare observed, "methinks they do protest too much", because people were legitimately challenging the science on which the anthropogenic global warming (AGW) hypothesis was based. Why was the reaction so extreme against legitimate scientific dissent? The answer is simple. The government and the agencies they set up through the UN, and specifically the Intergovernmental Panel on Climate Change (IPCC), were covering up a crime, and as we learned from Watergate, culpability is exposed in the cover-up. The Watergate scandal was a break in to the Democratic headquarters building of that name by agents of Richard Nixon's White House. That in itself was outrageous, but it was the attempts to cover up or deny that it happened that caused and compounded the greater scandal. The corrupted science of the IPCC was essentially controlled by a small group of scientists at the Climatic Research Unit (CRU). What they were doing was exposed in the surreptitious release of over 6000 emails from the Climatic Research Unit (CRU) at the University of East Anglia (UEA) in England that became known as Climategate.

This book is a challenge because it is the complexity of climate and science that the perpetrators, whom I will list in the next section of the book, exploited to bamboozle and deceive the world. It is difficult because people think weather and climate are simple concepts. It is as implied in the quote, "Everybody talks about the weather but nobody does anything about it." Most attribute it to Mark Twain, but it was written by Charles Dudley Warner, contemporary of Twain. Actually, people are always trying to do something to influence the weather; some include innocuous rain-dancing, others quite brutal in the sacrificing of children in South America among Pre-Columbian cultures. Precipitation was more critical to these cultures than most because they lived in an area that fringed the Atacama Desert, one of the driest places on Earth. Their priests observed the Pleiades star formation in the spring from high in the Andes to study the optical conditions. They knew empirically that the difference between a clear or shimmering cluster of stars determined the precipitation pattern. It was a useful rainfall predictor that guided them when to plant their main crop: potatoes. What they were actually witnessing was the influence of the El Nino pattern which determines precipitation patterns in that part of the world. As the global pattern of climate changed the beginning in about 1000 A.D., the pattern became less reliable and droughts more frequent. It is likely they adopted more extreme measures to appease the Gods out of desperation. Gradually, other techniques like cloud seeding, became more sophisticated, but not necessarily more effective.

Nowadays, governments plan or carry out all sorts of schemes collectively known as geo-engineering. As the web site geongineeringwatch. org says:

> *The absolute epitome of human insanity is the ongoing decades long attempt to completely engineer Earth's climate system (with countless variations of weather and biological warfare along the way). Even the U.S. National Academy of Sciences is trying to sound the alarm on dangers of geoengineering (though these scientists have not yet shown the courage to admit global geoengineering has long since been fully deployed).*

For example, the Canadian government was involved in spreading iron filings across the Pacific Ocean to try and increase absorption of

CO2 without success. They gave no thought to the damage done to the chemistry of the surface water. The US government sprayed chemicals at high altitude creating "Chemtrails" in order to reduce sunlight and offset global warming. Arrogantly and ignorantly, some scientists believe and try to convince governments that they can stop climate change. The truth is, if you don't understand the mechanisms or what is going on you are safer to do nothing because the chances of creating greater disaster is very high. We don't understand the mechanisms or what is going on. If we did, we could make accurate weather and climate forecasts, and we can't.

Why?

WHAT IS NORMAL?

The major theme of the claim that humans are causing global warming is that, as a result, current weather and climate patterns are not normal. The trouble is even a cursory look at historical weather records show they are well within normal patterns.

Weather is the atmospheric conditions at a place and time. It is the total of everything from cosmic radiation from space to heat from the bottom of the ocean and everything in between. Climate is the average of the weather over time or in a region. It is a statistic. The second part of the theme claims current weather and climate is not normal or not natural. For example, the 1999 *Greenpeace Report on Global Warming* says that carbon dioxide (CO_2) is added to the atmosphere naturally and unnaturally. But people read this without thought to the implications. What is "unnatural"? In this case, it is the human produced CO_2. If what humans produce is unnatural then by implication humans are unnatural. German philosopher Johann Wolfgan (von) Goethe provided the complete answer when he wrote, "The unnatural – that too is natural."

The simple way the public are fooled is that natural events are presented as unnatural. It works because people don't know the facts or the science. A classic example occurred recently. In July 2016 the UK Daily Mail newspaper illustrated the point

> *Melting in the Arctic reached an all-time high in June:*
> *Ice has been disappearing at a rate of 29,000 square*
> *miles a day.*

This is near the average daily rate of melt in the brief Arctic summer, but few people know this is natural. Approximately 10 million km2 of ice melts every summer in approximately 145 days, which is a melt rate of 68,965 km2 (26,627 square miles) per day.

Al Gore the idea that the current climate is normal in his documentary *An Inconvenient Truth* and in doing so showed either a lack of understanding or deliberate misdirection. For example, he does not use the word "normal" at all in the movie and uses the word "weather" twice and "change" 16 times. However, he implies normality by using the Goldilocks principle to say that the Earth's climate was ideal right now for humans being not too hot, not too cold, but just right. But that is his opinion today. People living during the last Ice Age 20,000 years ago would have considered their climate normal as did people living during the much warmer Medieval Warm Period (MWP) 1000 years ago.

Although Al Gore is a major target for much of this book, this is only because he chose the exploitation of environmental isues and global warming as the vehicle for his political ambitions. It is not coincidence that he shared his Nobel Prize with the IPCC. That organization and the people who created it are the power behind the throne. These people are mostly unknown to the public but are identified in the

"Who" section of this book. Without them, Gore would have none of the "pseudo-science," such as the sea level rise or arctic ice melting, to discuss in his movie.

To put the entire issue of "normal climate" in perspective, consider recent discussions about record weather events and, globally, the warmest year on record. The conclusion is totally misleading and scientifically useless. It is based on the very limited instrumental record that covers about 25% of the Earth's surface for, at most, 120 years. The age of the Earth is approximately 4.54 billion years, so the sample size is 0.000002643172%. Discussing the significance of anything in a 120-year record plays directly into the hands of those trying to say that the last 120 years of climate is abnormal and all due to human activity. This is the scientific equivalent of the standard political trick of taking information out of context. It is done purely for political propaganda, to narrow people's attention, and to generate fear. This point was made by Justice Burton of the UK Court who was asked to rule on the nature and objective of Gore's movie. In his judgment he wrote:

> *I viewed the film at the parties' request. Although I can only express an opinion as a viewer rather than as a judge, it is plainly, as witnessed by the fact that it received an Oscar this year for best documentary film, a powerful, dramatically presented and highly professionally produced film. It is built round the charismatic presence of the ex-Vice-President, Al Gore, whose crusade it now is to persuade the world of the dangers of climate change caused by global warming. It is now common ground that it is not simply a science film – although, it is clear that it is based substantially on scientific research and opinion – but that it is a political film, albeit of course not party political. Its theme is not merely the fact that there is global warming, and that there is a powerful case that*

such global warming is caused by man, but that urgent,
and if necessary expensive and inconvenient, steps must
be taken to counter it, many of which are spelt out.

Because people don't know what is normal regarding climate, they don't know that graph data can be picked to support a position. *The Guardian* newspaper headline for 24 March 2014 said, "13 of 14 warmest years on record occurred in 21st century – UN."

Figure 1 illustrates the situation for the period of the instrumental record starting in 1880. Of course, the highest temperatures occur in the most recent portion.

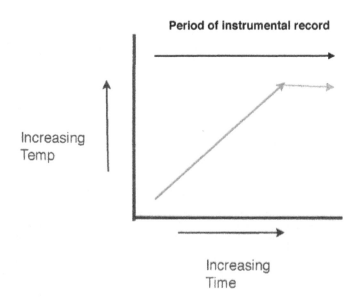

Figure 2 shows the same graph as Figure 1, but with the pre-instrumental temperature extended back to 1680, the nadir of the LIA.

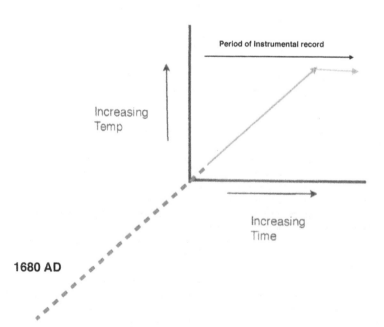

Figure 3 shows global surface temperatures from 1979, illustrating the problem that is not supposed to exist. From 1998 onward, the temperature shows a slight decline, but over the same period, global CO2 levels continued to rise. All the "official" computer model predictions of the IPCC, show temperatures rising over this period. Every one of their predictions has been wrong, which means the science is wrong, but instead of admitting that, they simply move the goal posts. I added a line at 1998 showing that before that year they talked about global warming. Afterward, it became climate change, because reality didn't fit their hypothesis. They get away with this because people don't understand the science or what is normal.

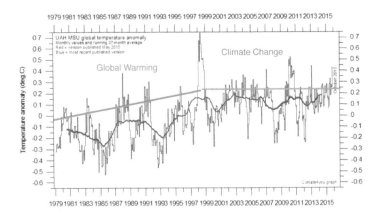

Governments say today's climate change is not normal when it is well within historical norms. They claim humans are causing global warming, but they established what is normal by cherry picking, omitting, or creating information, which has allowed them to say that any change is abnormal. It also means that every day they have another normal event to exploit. For example, a hurricane is normal, and the numbers vary in intensity and frequency throughout history, but that is not what people are told.

Alarmism exploits the philosophy of uniformitarianism – the idea that change is gradual over long periods of time. It evolved from Darwin's need to have an older world, with time for evolution to occur. It also helped make Darwin's theory of evolution palatable to an angry society. It created the false basis of the Western view of nature as being one of a gradual changing condition. This allows alarmists and promoters of AGW to say that any dramatic change is abnormal and must be caused by humans. This is false. Change is constant, dramatic, and can happen in very short time periods. Since 1900, global temperatures have gone through four trends of warming and cooling. Overall, the official records show that from 1900 to 1940 global temperatures rose while human production of CO2 was low; from 1940 to 1980 they fell while human CO2 production increased the most; from 1980 to 1998 they rose, and since 1997 they have fallen while human production of CO2 continues to increase.

97% CONSENSUS (THE ACTUAL # IS 0.3%)

From the beginning of the deception about AGW, there was a claim that a consensus of scientists agreed that humans were causing global warming. The claim was political because there was no consensus in science. As Einstein explained, "No amount of experimentation can ever prove me right: a single experiment can prove me wrong."

Two 97% claims were completely concocted. Gore used one claim in his movie, *An Inconvenient Truth*. It was the claim by Dr. Naomi Oreske, a science historian that with an internet search she found that 97% of articles were about human-caused global warming. The results of the search were predetermined by the keywords used:

Dr. Peiser used "global climate change" as a search term and found 1,117 documents using this term, of which 929 were articles and only 905 also had abstracts. Therefore, it is not clear which were the 928 "abstracts" mentioned by Oreskes, and Science did not, as it would have done with a peer-reviewed scientific paper, list the references to each of the "abstracts". Significantly, Oreskes' essay does not state how many of the 928 papers explicitly endorsed her very limited definition of "consensus". Dr. Peiser found that only 13 of the 1,117 documents – a mere 1% – explicitly endorse the consensus, even in her limited definition.

Undoubtedly, the most devastating comment about Oreske's work was made by Tom Wigley, the former Director of the CRU who became the center for the corruption as exposed by leaked emails, and grandfather to the corrupted climate process. Commenting on a work similar in shallowness he wrote, "Analyses like these by people who don't know the field are useless. A good example is Naomi Oreske's work."

The next deception, also about 97 percent, was perpetrated by Australian academic John Cook. Cook claimed that 97 percent of 11,944 abstracts agreed that "Human activity is very likely causing most of the current global warming".

What wasn't disclosed is that only 4,014 of the 11,944 abstracts even expressed an opinion and of those only 41 fit the author's definition of totally agreeing. So the actual result was 0.3% agreed.

The research was concocted to achieve a goal. It was frighteningly successful because few know, even now, that it is completely false in creation and conclusion. Despite this, it is a powerful comment when used by President Obama from the bully-pulpit.

PARADIGM SHIFT: ENVIRONMENTALISM

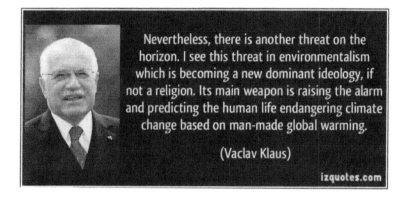

Nevertheless, there is another threat on the horizon. I see this threat in environmentalism which is becoming a new dominant ideology, if not a religion. Its main weapon is raising the alarm and predicting the human life endangering climate change based on man-made global warming.

(Vaclav Klaus)

izquotes.com

A paradigm shift is a significant change in the way society views something. Most people are slow to accept shifts because they fear change. They know that with change, some get hurt but some benefit. Most fear they might be the ones hurt. However, there is always a small group who grab the paradigm shift and exploit it for their agenda, either financially or politically or both. That is what happened with environmentalism. We are all environmentalists, but a few grabbed it and claimed they were the only ones who cared about the environment and preached to the rest of us from that moral high ground. The damage that was done to society, particularly through energy policies, is now appearing.

Most see the good ideas in the 20th century as new paradigms such as feminism or environmentalism; the problem is that most people usually don't know how much of it to adopt. Things occur with no clear purpose. For example, what is the role of extremism? I realized by studying the impact of paradigm shifts on society that extremists define the limits for the majority. The comment, "well, now they are going too far" encapsulates it. Hopefully, this realization occurs before too much damage is done. Environmentalism is a new and completely necessary paradigm. It makes no sense to soil your nest, but a few grabbed the paradigm and used it to bully others by claiming only they cared. The majority of people care and embraced the basic ideas. The problem most always have is how far do we take the idea. People are justifiably afraid of change because they know there are always winners and losers. With limited control over their lives and restricted abilities to adjust, they are afraid they will lose.

For a long time, I wondered about the role of extremism in society. Gradually, I realized that they defined the limits of issues for the majority. Although changes are still needed, the extremists made the majority realize that they were losing more than they were gaining. Extreme feminists performed that function for the majority of women. As environmentalists and global warming activists lose the public debate, they become more extreme. A few cold winters in the first decade of the 21st century caused a problem. AGW proponents claimed the cold was due to warming. It was so illogical that even people who didn't understand the science knew something was fundamentally wrong. Similarly, many, including some hardened liberal journalists, knew Al Gore was wrong when in 2007 he told a joint session of the House Energy Committee and The Senate Environment Committee that the climate debate was over, the science was settled. The journalists knew, as any moderately informed person does, that science is never settled. Now, as more and more people learn what was done and that the forecasts are wrong, the AGW proponents, instead of acknowledging that their hypothesis was wrong, have chosen to make more extreme claims. This is further proof that the AGW hypothesis, and its promotion, is for a political agenda. This growing concern is a threat to the credibility of science and environmentalism, a classic example of crying wolf. When people learn how much they were deceived, they will say they don't believe anything, and environmentalism will lose momentum.

EXPLOITATION OF COMPLEXITY

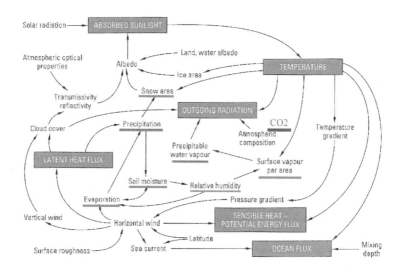

The diagram shows a simple systems display of the atmosphere. It illustrates how complex it is and, therefore, how difficult it is to reduce it to a computer model. Despite this, those who pushed the human-caused global warming argument focused on one fractionally small part – CO2 , a part of the segment labeled *"atmospheric composition"*. AGW advocates the claim that human CO2 is almost the sole cause of global warming and climate change. In fact, CO2 is about 4% of the total greenhouse gases that comprise atmospheric composition. Water vapor (water in gaseous form) is 95% of the greenhouse gases by volume, essentially considered constant, but this is ignored. The IPCC assumed that the amount of water vapor humans produce is of no consequence relative to the total volume. The problem is the natural variability of water vapor in the atmosphere far exceeds any greenhouse effect the fraction of human produced CO2 may have. Of course, this assumes we have sufficiently accurate measures of atmospheric water vapor, which we don't.

As recently as 2000 we learned that:

> *After years of sustained research efforts into the accuracy of atmospheric water vapor measurements, researchers from the U.S. Department of Energy's ARM Program have succeeded in reducing measurement uncertainties from greater than 25% to less than 3%.*

This sounds promising, but it is only a reduction of uncertainties, and provides no historic evidence of variability.

Also, the focus is on temperature and only on warming, but for life on Earth in the short term, precipitation is more important. In the diagram, I underlined in red all variables affected by water and water vapor. We have virtually no measures of any of these variables. For example, the official UN climate agency, the Intergovernmental Panel on Climate Change (IPCC), involved in establishing climate science admits we have virtually no measures of soil moisture. The lack of precision and complete inadequacy of the data should preclude any scientific results obtained from such data being used as the basis for global climate and energy policies. Of course, that hasn't happened, because it is for a political agenda not science. As Michael Crichton said at the end of his speech,

> *Because in the end, science offers us the only way out of politics. And if we allow science to become politicized, then we are lost. We will enter the Internet version of the dark ages, an era of shifting fears and wild prejudices, transmitted to people who don't know any better. That's not a good future for the human race. That's our past. So it's time to abandon the religion of environmentalism,*

*and return to the science of environmentalism, and base
our public policy decisions firmly on that.*

MOVING GOALPOSTS

*Global Warming became Climate Change became
Climate Chaos became Climate Disruption*

An indication of the use of climate for a political agenda is the constant shifting of terminology. It began in 1988 as the threat of global warming. That date is specific because James Hansen of NASA Goddard Institute for Space Studies (GISS) was chosen to appear before the Senate committee; he was willing to say he was 99% sure that humans were causing global warming.

As Senator Wirth explained in a 2007 *PBS Frontline* interview:

> *... I don't remember exactly where the data came from,
> but we knew there was this scientist at NASA who had*

really identified the human impact before anybody else had done so and was very certain about it. So we called him up and asked him if he would testify. Now, this is a tough thing for a scientist to do when you're going to make such an outspoken statement as this and you're part of the federal bureaucracy. Jim Hansen has always been a very brave and outspoken individual.

Plucked from obscurity to give the testimony the politicians wanted, Hansen was rewarded with becoming Director of NASA's Goddard Institute of Space Studies (GISS). He used this agency to advance the AGW hypothesis by various nefarious means. More important, he openly defied the Hatch Act introduced by Congress in 1939 to set limitations on bureaucratic involvement in politics. He was arrested on a few occasions protesting against coal plants, including one occasion outside the White House. He was the personification of bureaucratic control and promotion of the deception of global warming.

Around 1998, global temperatures began to level and decline while CO2 levels continued to increase. This was in complete contradiction to the AGW hypothesis and indicated the science was wrong. Correct science re-examines the hypothesis. Incorrect science, as practiced, simply changed the name from global warming to climate change. People realized climate always changes, so, John Holdren introduced the term Climate Disruption. It is effective because it implies abnormality and, therefore, due to humans.

It is not necessary to cover climate change for the entire history of the Earth to illustrate how much it changes naturally. It is necessary to show that current climate is well within natural variability and, contrary to the propaganda, that we are not at the warmest ever.

LACK OF KNOWLEDGE OF HISTORY OF CLIMATE CHANGE

The claim is that the world is currently warmer than it has ever been. This diagram shows the interglacial or warm periods between glaciations over the last 420,000 years. It is part of the period known as the Pleistocene, the most recent glacial period. Within the Pleistocene, ice formed and advanced and retreated in a series of glacial and interglacial events. Notice that the current interglacial is lower in temperature than any previous interglacial, disproving the claim it is warmer now than ever before. The diagram illustrates how much temperature changes naturally all of the time. The temperature range in the diagram is some 12°C. It also shows the cyclic nature of temperature.

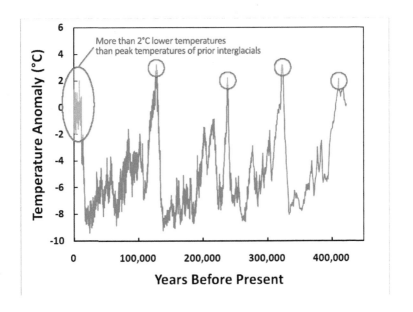

ICE SHEETS JUST 18,000 YEARS AGO

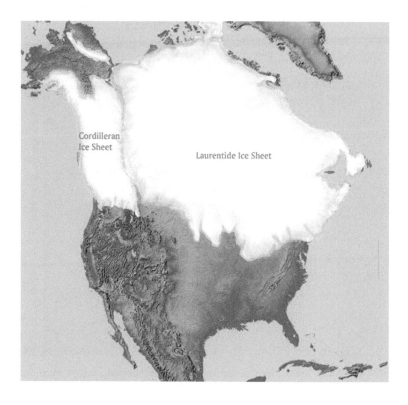

Cordilleran
Ice Sheet

Laurentide Ice Sheet

The left diagram shows the world 20,000 years ago looking down on the north pole with vast ice sheets stretching across the northern hemisphere. The right diagram shows two large ice sheets, with a combined area larger than the current Antarctica sheet covering North America. Water to create the snow and ice came from the oceans. As a result, the sea level was lower than today's just 20,000 years ago. Most of the ice melted very rapidly in about 7000 years.

SEA LEVEL CHANGE

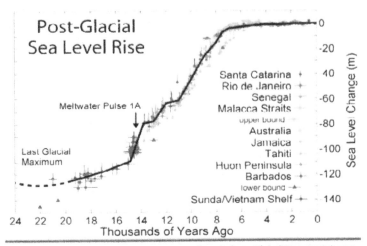

Image from GlobalWarmingArt.com

The diagram above shows the sea level from 18,000 years ago at 130 m (427 feet) lower than at present. Because the ice melted in about 7000 years, (15,000 to 8000 years ago), sea levels rose close to the current level about 7000 years ago. Since that time, there has been very little increasing sea level, contrary to what the media reports, primarily because of Al Gore's disgracefully inaccurate movie, *An Inconvenient Truth*. Notice the dramatic rise identified as Meltwater Pulse 1A, during which the sea level rose 20m in less than 500 years, and some claim 200 years. The point is, the scares about the current sea level rise, being unheard of and threatening, are, like the temperature and severe weather claims, false. Like everything else, the claims are based on inadequate data and subjected to selective data selection. As climate expert Patrick Michaels notes:,

> *The IPCC claims a faster rate in sea level rise in the period 1993-2003 (3.1 mm/year) compared with*

1961-2003 (1.8 mm/yr). See WG1 SPM p 5,7, table SPM1. To make this claim, the IPCC has employed two of their familiar misleading tricks simultaneously –- (a) compare a short period with a longer period, (b) change the measurement technique.

NORTHERN HEMISPHERE TEMPERATURE FOR LAST 10,000 YEARS – THE HOLOCENE OPTIMUM

Greenland GISP2 Ice Core - Temperature Last 10,000 Years

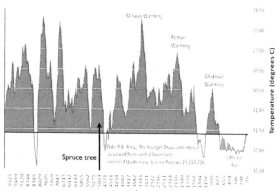

White Spruce (Picea Glauca)
Radio carbon date 4940 ±140
Located 100 km north of current tree line
Source: Photo by Professor Ritchie in H H Lamb, Vol. 2 1972

Warmer temperatures caused the rapid ice melt. This diagram illustrates how the temperatures for the last 10,000 years, a period known as the Holocene Optimum, were almost all warmer than at present. The photograph shows a fossilized white spruce (Picea Glauca) 100 km north of the current tree-line and radiocarbon dated at approximately 5000 years old. Global temperatures were 2°C warmer than at present for the tree to exist at that latitude. This stark hard evidence completely contradicts Gore's claim that the world is warmer than it has ever been.

THE INSTRUMENTAL GLOBAL TEMPERATURE RECORD

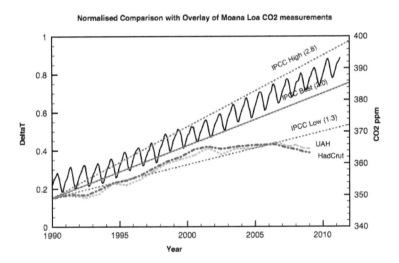

Normalised Comparison with Overlay of Moana Loa CO2 measurements

Thomas Huxley, 1825-1895, biologist ardent advocate for Darwin, said the great tragedy of science was a lovely hypothesis destroyed by an ugly fact. This diagram shows temperatures since 1990. They rose to 1998 but have leveled and declined since. More problematic for the IPCC agenda was that CO2 levels continued to rise over this same period. This contradicts their hypothesis that a CO2 increase because of human industry would cause a temperature increase; it was an ugly fact and untrue.

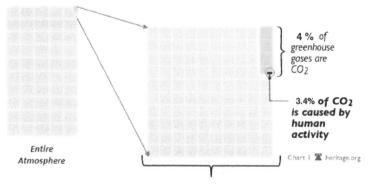

4 % of greenhouse gases are CO_2

3.4% of CO_2 is caused by human activity

Entire Atmosphere

Chart I ☎ heritage.org

This block represents all greenhouse gases, which comprise only 2% of the total atmosphere

We need some facts about the natural conditions of CO2 to understand the real picture. Figure 4 is a schematic designed to give perspective and context to the various percentages of greenhouse gases.

1. The IPCC assumes an increase in CO2 causes a temperature increase. It doesn't. Every record for any period shows that temperature increases *before* CO2. The only place a CO2 increase precedes a temperature increase is in IPCC computer models.

2. The amount of CO2 was inconveniently small relative to H2O (water vapor), so they concocted an effectiveness value. They said each CO2 molecule reduced the escape of long wave (LW) radiation at a higher level than a molecule of water vapor. The trouble is the range of estimates varies considerably. If this effect follows the laws of physics, then all calculations should give the same results.

3. Let's assume CO2 is causing warming. Then, when its density reaches a certain level, the warming ability is maximized. It is currently maximized, so the addition of more CO2 has little effect. Consider light passing through a window. One coat of black

paint blocks most of the light, but subsequent coats block only fractionally more. Doubling or tripling atmospheric CO_2 has little further temperature effect. To bypass the problem, IPCC theorized a positive feedback. Higher temperature due to CO_2 increases evaporation and more water vapor causes increased temperature. It doesn't happen. In fact, the feedback is negative partly because of increased cloud cover.

4. The IPCC says the current level of atmospheric CO_2 is 400 parts per million (ppm). Al Gore claims this is the highest ever. Actually, it's the lowest in 600 million years. We know plants function best at 1200 ppm because commercial greenhouses inject that level, which was the average level of the last 300 million years.

5. Statistics on the annual human production levels of CO_2 are produced by the IPCC. In an FAQ section of their Reports, the IPCC asks the question, "How does the IPCC produce its inventory Guidelines?" The answer:

> *Utilizing IPCC procedures, nominated experts from around the world draft the reports that are then extensively reviewed twice before approval by the IPCC. This process ensures that the widest possible range of views are incorporated into the documents.*

What it ensures is they have control of the numbers and make sure they constantly increase. The problem is with the amount of CO_2 they claim humans produce annually – about 10 gigatons, which is within the error range of estimates of two natural sources: oceans (90-100 gigatons), and soil bacteria/decomposition (50-60 gigatons).

6. The IPCC claim the pre-industrial level was 270 ppm. Approximately 90,000 measures of atmospheric CO2 that began in 1812 have showed pre-industrial levels of 335 ppm. Ice core expert Zbigniew Jaworowski explains, *"The basis of most of the IPCC conclusions on anthropogenic causes and on projections of climatic change is the assumption of low level of CO2 in the pre-industrial atmosphere. This assumption, based on glaciological studies, is false."*

All official atmospheric levels of CO2 are measured by a system created on Mauna Loa, a volcano on the Island of Hawaii, patented and owned by the Keeling family. Ernst-Georg Beck, co-founder of the European Institute for Climate and Energy (EIKE), explained that Charles Keeling established the readings by using lowest afternoon readings while ignoring natural sources. Beck notes, "Mauna Loa does not represent the typical atmospheric CO2 on different global locations but is typical only for this volcano at a maritime location in about 4000 m altitude at that latitude." Keeling's son now operates Mauna Loa and as Beck notes, "owns the global monopoly of calibration of all CO2 measurements." He's also a co-author of the IPCC reports.

Beck contacted me when he began work on the 90,000 CO2 measures from the 19th century. They were taken following discovery of oxygen that triggered the desire to identify all the atmospheric gases. I warned Beck about the attacks he would endure because "the establishment" had manipulated the data to claim the low pre-industrial level of CO2. In an obituary, a long-time friend of Beck wrote:

> *Due to his immense specialized knowledge and his methodical severity, Ernst very promptly noticed numerous inconsistencies in the statements of the*

Intergovernmental Panel on Climate Change (IPCC). He considered the warming of the earth's atmosphere as a result of a rise of the carbon dioxide content of the air of approximately 0.03 to 0.04 percent as impossible. And he doubted that the curve of the CO2 increase noted on the Hawaii volcano Mauna Loa since 1957/58 could be extrapolated linearly back to the 19th century. (Translated from German)

On November 6, 2009 Beck wrote to me about personal attacks saying, "In Germany, the situation is comparable to the times of the medieval inquisition."

7. 20th century temperatures increased most from 1900 to 1940 with little increase in human production of CO2. Human production of CO2 increased most from 1940 to 1980, and total CO2 levels increased, but temperatures went *down* as Figure 5 shows.

8. IPCC assumes CO2 is uniformly distributed in the atmosphere. It is not; it varies considerably as recent satellite imagery shows.

Since 1998, CO2 levels increased, but temperatures declined in contradiction to the IPCC assumption. IPCC predictions were wrong because their CO2 facts and science are wrong. Global temperatures are declining. The data in Figure 6, prepared by aeronautical engineer Burt Rutan, uses IPCC data to show the cooling.

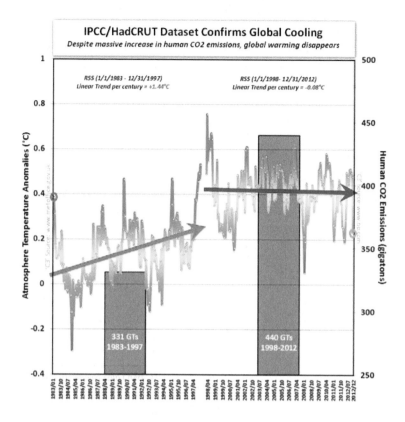

The diagram compares CO2 and temperature trends for two periods. It illustrates the contradiction to the IPCC claim that asserted with 90+ percent certainty that human produced CO2 is the cause. The diagram shows that the decline in temperature since 1998 occurred while human production increased more than in the period before 1998. The period without warming is called the pause, or hiatus.

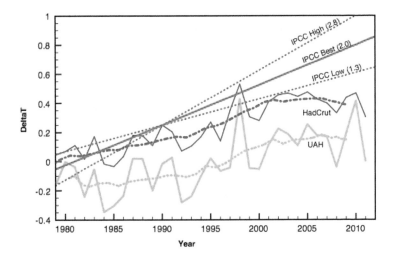

Cartoonists always capture and concentrate the illogical. This one shows 17 years; it is now 21 and counting.

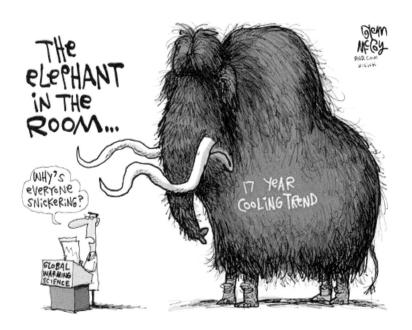

IPCC TEMPERATURE PROJECTIONS VERSUS REALITY

Figure 7 compares IPCC computer model projections (red lines) for temperature starting in 1990 against the reality as measured by two temperature records, HadCrut (surface data) and UAH (satellite data).

Actually, IPCC predictions were already wrong by 1995, so they stopped calling them predictions and started calling them projections. Notice that even the lowest "projection" is incorrect.

COMPUTER MODELS: GOSPEL IN GOSPEL OUT (GIGO)

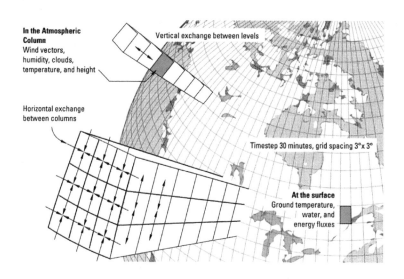

Figure 8 shows the basic global divisions created for a computer climate model. They are cuboid, based on a grid system measured by latitude and longitude and a vertical scale. The claim is the smaller the grid, the more accurate the model, but this is inaccurate because it doesn't matter how fine the grid is if the data is not available. Even the smallest grid size is so large that major weather systems such as thunderstorms and tornadoes cannot be included.

There is virtually no weather data for some 85 percent of the world's surface. Virtually none for the 70 percent that is oceans, and of the remaining 30 percent land, there are very few stations for the approximately 19 percent mountains, 20 percent desert, 20 percent boreal forest, the 20 percent grasslands, and 6 percent tropical rain forest. The horizontal and vertical global coverage is inadequate, but it also lacks the length of record for adequate model construction. There only 1000 stations with 100 years of record and they are almost all in the Eastern US and Western Europe.

H. H. Lamb, founder of the Climatic Research Unit (CRU), recognized the problem and said it was the reason for forming the Unit:

> ... it was clear that the first and greatest need was to establish the facts of the past record of the natural climate in times before any side effects of human activities could well be important.

There are ways of estimating past temperatures, but the degree of accuracy diminishes very quickly. Indeed, even the modern record is very inaccurate.

The 2001 IPCC Report, using data prepared by Phil Jones, Director of the CRU, said the global temperature average, reportedly using the best modern instrumental database over the longest period of data available, rose 0.6°C over 100+ years. The problem is the error factor was ±0.2°C or ±33.3%. So, the modern instrumental temperature record, which is supposedly many times more accurate than any paleoclimate temperature record, is useless. Compare the Jones number of temperature change in a 100+ record with the difference between GISS and HadCRUT (surface data) in any given year. If, for the sake of argument, the difference is 0.1°C, then it is one-sixth of the difference for the claimed total change in 100+ years. This underscores the enormity of the error factor that makes the data essentially meaningless, especially as justification for draconian global energy and environment policies.

Models are three-dimensional mathematical cuboid representations of the atmosphere and oceans. There are built on data, but the surface data is completely inadequate.

It's worse in the vertical, with virtually no data in space and time, and constantly changing very complex conditions. Again, the illusion exists that increasing the number of layers built into the model creates better results. There are virtually no weather data measures, and even fewer measures of atmospheric composition such as changing aerosol types and volumes.

There is inadequate data for building computer models and partly because of that, the forecasts are always wrong. To get the results the political masters wanted, official data agencies, such as NASA GISS, altered all the records.

IPCC COMPUTER MODELS: AVERAGE OF AVERAGES

Climate models produce a different result every time they are run, so each model is run a few times, and the results averaged. The IPCC create an ensemble of 22 different models run by different agencies and countries, then they average the results; the final is an average of averages.

COURTING DISASTER
(THE CASE FOR AGW)

INABILITY TO FORECAST
WEATHER AND CLIMATE

Weather Forecast for Thu, Oct 23, 2014, issued 3:22 AM EDT
DOC/NOAA/NWS/NCEP/Weather Prediction Center
Prepared by Mcreynolds based on WPC, SPC and NHC forecasts

Weather forecasts have made virtually no improvements since Lavoisier (1743-1794) said over 200 years ago, "It is almost possible to predict one or two days in advance, within a rather broad range of probability what the weather is going to be." Today, anything beyond 48 hours grows increasingly inaccurate, and the performance is even worse for severe weather.

AGW proponents argue that weather forecasts are different than climate forecasts. They are not. The problem is climate is the average of the weather. Medium and long-range climate forecasts are also failures. In 2008, Tim Palmer, a leading climate modeler at the European Centre for Medium-Range Weather Forecasts in Reading England, said in the New Scientist, "I don't want to undermine the IPCC, but the forecasts, especially for regional climate change, are immensely uncertain." The comment applies to medium length forecasts but is equally true for short and long term forecasts.

Professor John Christy of the University of Alabama at Huntsville (UAH) in a presentation before Congress showed the graph of IPCC temperature forecasts. The red line is the average of 102 different models plotted against the actual temperatures shown in blue and green squares. Every single prediction, they call them projections, since the IPCC inception in 1990, is wrong. Professor Richard Feynman, Nobel Laureate in Physics, said, *"It doesn't matter how beautiful your theory is, it doesn't matter how smart you are. If it doesn't agree with experiment, it's wrong."* In other words, if your prediction is wrong your science is wrong.

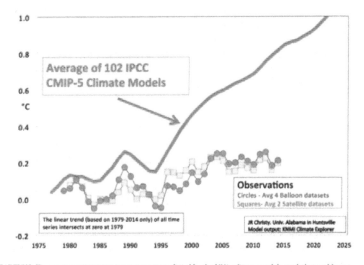

CAPTION: Five-year running mean temperatures predicted by the UN's climate models, and observed lower atmospheric temperatures from weather balloons and satellites.

EXPLOITATION OF PUBLIC LACK OF KNOWLEDGE AND EMOTIONS

Eco-bullying is just another form of modern day bullying. It works because so few people know anything about the way the Earth works. I know, after teaching a science credit for arts students and giving hundreds of public lectures over the years.

Science has become so specialized that even many scientists don't understand. Change and dramatic change are the norms. It is easier when you can present natural weather events as unnatural. The goal is to get media attention and a headline. The headline is usually definitive and in the active voice. The actual story usually has all sorts of conditional phrases, such as "It appears that...", and "it could happen...". This effectively allows them a disclaimer, if the information is later exposed as incorrect.

MEASURES OF PUBLIC CLIMATE KNOWLEDGE

In a 2010 study by Yale University, of public knowledge of climate, over 50 percent of students got a failing grade. What is even more telling is how the researchers responded to the results. They decided the poor results were their problem by having asked difficult questions. So, as is done in education to cover teaching incompetence, they applied a "curve". Here are the results.

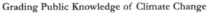

Grading Public Knowledge of Climate Change

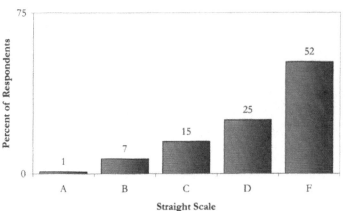

The researchers did what too many schools do, they changed the results. It is hard to understand why. They decided the questions were too hard, and "curved" the results. It is a reflection of the growing dishonesty in academia, generally, and their research specifically. If the questions were too hard, then they didn't know what they were doing. Presumably they altered the results to produce a favourable outcome to satisfy the funding agency. From a total of 77 percent getting D or F, they improved it to 73 percent getting A, B, or C.

Grading Public Knowledge of Climate Change

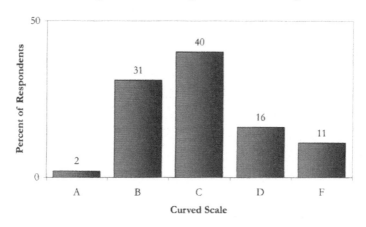

Curved Scale

CLIMATEGATE

The few people involved in climatology before 1990 knew what the IPCC was doing from the start. There were so few actually studying climatology that they were easily marginalized by personal attacks and lawsuits. Marginalization involved creating a group, so it was a fringe. First, it was Global Warming Skeptics, knowing the public view of skeptics is different and that all scientists must be skeptics. When the facts changed, the concern switched; Skeptics became Climate Change Deniers, with the Jewish holocaust connotation.

The corruption of science continued, mostly at the Climatic Research Unit (CRU) at the University of East Anglia (UEA). Members controlled important chapters like data, computer models, and paleoclimate information. Maurice Strong, creator of the United Nations Environment Program (UNEP) and, as we will see later, the mastermind behind the entire global warming deception, consolidated the political side through the Conference of the Parties (COP), which made political decisions based on IPCC science. Initial focus involved the Kyoto Protocol, designed to punish developed nations for creating the CO_2 that was causing damaging global warming. The claim is that the human portion of CO_2 is identified by the lower levels of C14, which doesn't exist in fossil fuels. This assumes there is no increase in the natural fires of coal seams and peat across the world. But allowing for that, the amount of human production is miniscule and it assumes CO_2 is a greenhouse gas. They would pay developing nations compensation, in a massive transfer of wealth.

Kyoto was in political trouble, but the COP was determined to save it at the conference scheduled for Copenhagen in December 2009 (COP 15). Then somebody leaked 1000 emails from the CRU, who disclosed a series of wrongdoings, all designed to achieve the climate results they wanted, namely that human created CO_2 was the cause. They controlled peer reviews by reviewing each other's work; they attacked editors including getting one fired; and created a website to attack and create propaganda. One person, William Connolley, controlled over 500 Wikipedia items, until he was exposed. A year later, another 5000 emails were released, detailing the degree of manipulation, corruption, and control. The COP acts on the basis of the IPCC science, so the leaked emails exposed the corrupted science and killed the Kyoto Protocol.

They didn't wait long. A year later at COP16 in Durban South Africa, they created the replacement Green Climate Fund that was ratified at COP 21 in Paris in 2015.

On the cover of their 2010 book *Climategate,* Mosher and Fuller summarized what the leaked emails reveal and it is the very antithesis of proper science.

> *"The Team led by Phil Jones and Michael Mann, in attempts to shape the debate and influence public policy:*

- *Actively worked to evade McIntyre's Freedom of Information requests, deleting emails, documents, and even climate data*

- *Tried to corrupt the peer-review principles that are the mainstay of modern science, reviewing each other's work, sabotaging efforts of opponents trying to publish their own work, and threatening editors of journals who didn't bow to their demands*

- *Changed the shape of their own data in materials shown to politicians charged with changing the shape of our world, 'hiding the decline' that showed their data could not be trusted."*

COVER UP

Despite all this evidence from the scientists themselves a decision to pursue a cover-up was taken by major agencies directly involved. They initiated five inquiries. Clive Crook, senior editor of The Atlantic, wrote:

> I had hoped, not very confidently, that the various Climategate inquiries would be severe. This would have been a first step towards restoring confidence in the scientific consensus. But no, the reports make things worse. At best they are mealy-mouthed apologies; at worst they are patently incompetent and even wilfully wrong. The climate-science establishment, of which these inquiries have chosen to make themselves a part, seems entirely incapable of understanding, let alone repairing, the harm it has done to its own cause.

Here is a summary of the investigations:

There was a distinct pattern to the process used in each inquiry, which was clearly dictated by the objective of the cover up.

- The people appointed to the inquiries were either compromised through conflict or had little knowledge of climatology or the IPCC process.

- They did not have clearly defined objectives and failed to achieve any they publicized.

- Interviews were limited to the accused.

- Experts who knew what went on and how it was done, that is, understood what the emails were saying, but were not interviewed.

- The validity of the science and the results obtained as published in the IPCC reports were not examined, yet the deceptions were to cover these problems.

- All investigations were seriously inadequate in major portions so as to essentially negate their findings. It appears these inadequacies were deliberate to avoid exposure of the truth.

- They all examined only one limited side of the issues, so it was similar to hearing only half of a conversation and what is heard is preselected.

Those involved in the cover up achieved their goal because the media stopped asking questions. It also allowed those identified in the emails to claim they were absolved of any wrong-doing.

MISINFORMATION ABOUT SEVERE WEATHER

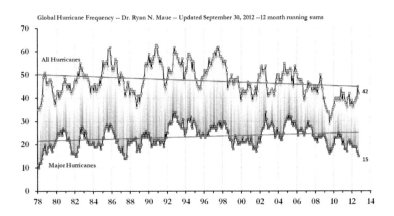

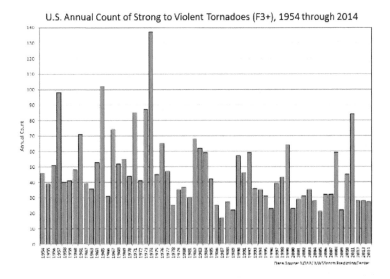

The public believes there is an increase in severe weather, such as hurricanes and tornadoes. There isn't. The plot of tropical storms and hurricane frequencies for the period 1971to 2012 (Figure 8), and tornadoes 1954 to 2012 (Figure 9), show that the increase is false.

To the general public, it appears that there are more because the media report on every little weather event. It is the same phenomenon of being introduced to a person and afterward, every time you turn around, there they are. They were always there, but just not a part of your awareness. It is the phenomenon that makes public relations (PR), some would say the modern term for deception or "spinning", effective.

Besides, in the case of global warming, the claim is based on incorrect science. Most severe weather occurs at the boundary between the warm tropical air and the cold polar air known as the Polar Front (Figure 9).

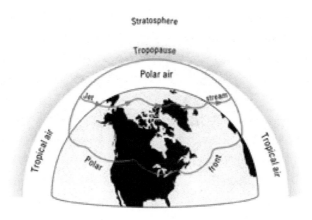

The number and severity of the storms along the Front are determined by the temperature difference across it. The greater the difference, the more severe the storms. The IPCC says this difference will decrease because the polar region will warm more than the tropical region.

ARCTIC ICE: A CLASSIC EXAMPLE
OF EXPLOITATION

When I ask people what is wrong with global warming, most can't answer right away. After a while, most say the sea level will rise, a major focus in Gore's movie, *An Inconvenient Truth*. Melting Arctic ice was something almost nobody understood then nor do they today.

ARCc0.08-04.1 Ice Concentration (%): 20160402

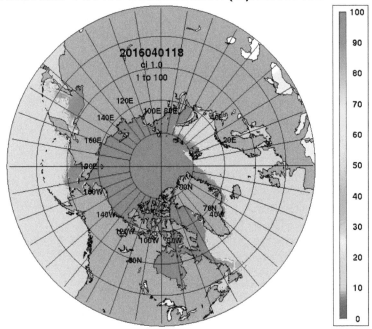

The diagram (Figure 10) is a polar projection (the North Pole is at the center), showing maximum ice extent for April 1, 2013. It shows a total area of ice of approximately 15 million square kilometres, which is almost twice the size of the United States. Few people have any idea because their view of the Arctic is a thin line across the top of a Mercator projection, a school wall map (Figure 11). In this projection, the North Pole, a single point, becomes a line as long as the Equator and the size and shape of the Arctic Ocean and Antarctica are unrecognizable. For example, how long did it take you to figure out the map of ice concentration above?

ARCc0.08-03.5 Ice Concentration: 20121012

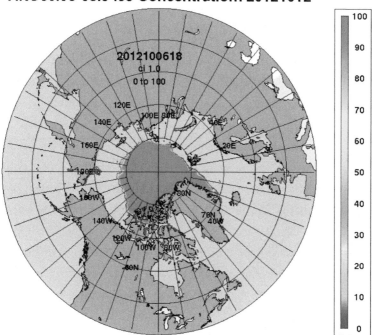

WHAT IS NORMAL FOR ANNUAL ARCTIC ICE MELT

The satellite image for May 1, 2012, shows maximum Arctic sea ice cover of approximately 15 million square kilometres (km2).

The second image for September 15, 2013, shows minimum ice cover of approximately 5 million km2. Each year, on average, approximately 10 million km2 melts in just 150 days. This is an area approximately equal to the continental US. It is a melt rate of 67,000 square kilometres *each day*, an area slightly larger than West Virginia or twice the size of Vancouver Island, **every single day.**

Accurate measures of Arctic sea ice began with a satellite in 1978. The actual data is only accurate since 1980 because it took time to calibrate and understand what they was being seen. Even with satellite images, different agencies get different results.

The world warmed from 1980 to 1998 with the result that the sea ice extent declined slightly. Based on this simple trend, some, including NASA, began to predict that summer ice would be completely gone by 2013. The decreasing trend of summer ice changed, starting in 2012. The same seasonal pattern of formation and melt of sea ice occurs round Antarctica and recently has set record amounts since 1980. Even the Arctic and Antarctic sea ice record of less than 40 years shows variability, that is, like all the very short modern records, well within natural variability. Consider what the ice conditions were like during the Medieval Warm Period or when that spruce tree was growing in a world at least 2°C warmer than today. How did the polar bears survive? How did they survive the 9000 years of warmer temperatures of the Holocene Optimumthan today?

YOU NEED A THREATENED ANIMAL
TO RAISE EMOTIONAL QUOTIENT

Al Gore chose polar bears, but they are not in decline. Read the work of zoologist Susan Crockford or the studies of biologist and polar bear expert Mitch Taylor who works for and with the Inuit people of Iglulikin, Nunavut. Inuit (Canadian word for Eskimo) leaders know the stories of decreasing polar bear numbers is false. Taylor wrote in a *Nunatsiaq News* article:

The Inuit have always insisted the bears' demise was greatly exaggerated by scientists doing projections based on fly-over counts, but their input was usually dismissed as the ramblings of self-interested hunters...The Inuit were right. There aren't just a few more bears. There are a hell of a lot more bears.

The tourist who took the polar bear image, shown above that Gore exploited, said he completely misrepresented the image. The polar bears climbed up on the ice to get a good view of the ship going by. Polar bears are superb open water swimmers. They survived many times when Arctic ice was reduced or even non-existent such as during the Holocene Optimum. They only recently (90-100,000 years ago) arrived in the arctic. They evolved from Alaskan Brown bears. The recent state of evolution is reflected in the natural occurrence of hybrids.

THE GREENHOUSE EFFECT

The Earth's atmosphere does not work like a greenhouse. It was a convenient analogy because it implies heat being trapped. It also identifies CO2 as an important gas to global warming, when it isn't. The key to the political greenhouse effect is the role of certain gases, called greenhouse gases (GHG), that keep the global temperature approximately 33°C warmer than without an atmosphere. Specifically, the Greenhouse Effect is the analogous claim that Earth's atmosphere functions like a greenhouse. It allows shortwave sunlight in, which heats surfaces producing long wave energy or heat, which is blocked completely from escaping by the glass. In fact, the analogy is incorrect; the atmosphere only acts like a greenhouse in a few simple ways. For example, the heat is moved in the atmosphere by evaporation, transport by wind, then released after condensation. This is a major atmospheric function that does not occur in the greenhouse.

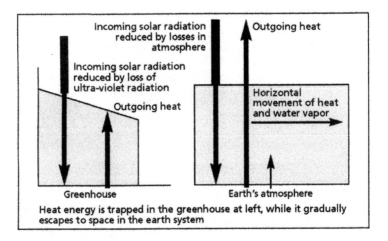

There are three gases that account for 99.9% of the GHG. They are water vapour (H2O), carbon dioxide (CO2) and methane (CH4).

The most important and most abundant by far, is water vapour at 95+ %, while CO2 is approximately 4%, and methane less than 0.01%. Of course, this assumes that CO2 is a GHG, which many people dispute. Its amount and effect are small – the human portion even less.

DELIBERATE DECEPTION OF PUBLIC PERCEPTION

The first pie chart shows the actual percentages of greenhouse gasses.

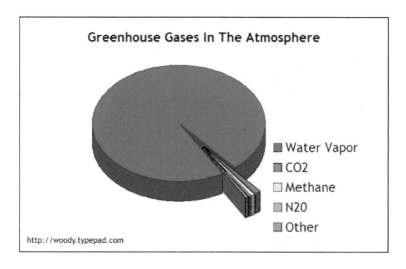

Greenhouse Gases In The Atmosphere

- Water Vapor
- CO2
- Methane
- N2O
- Other

http://woody.typepad.com

The next pie chart is completely wrong as the comment "Carbon dioxide is clearly the majority" attests. Nonetheless, it was on the ABC news web site as the label indicates.

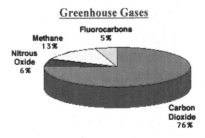

Greenhouse Gases

Methane
13%

Fluorocarbons
5%

Nitrous
Oxide
6%

Carbon
Dioxide
76%

This graph shows the distribution of GHG in Earth's atmosphere. Carbon Dioxide is clearly the majority.
www.abcnews.com/sections/us/global106.html

The claim is scientists are only measuring "dry air", but water vapour is so dominant, it is meaningless to say that. We should rename the planet, Water, because it is fundamental to everything, including life on the planet.

The bigger problem is that the IPCC did not follow the normal scientific method that requires the creation of a hypothesis based on a set of assumptions. The hypothesis was that a CO2 increase would cause a temperature increase, and CO2 will continue to increase because of human industrial and other activities. This became known as the anthropogenic global warming (AGW) hypothesis, discussed earlier. Normally, other scientists perform their role as skeptics, by challenging and trying to disprove the hypothesis. Instead, they worked to prove it and attacked any who challenged it. As Dr. Richard Lindzen said, the consensus was reached before the research had even begun.

Several actions were taken, as we will see later, but a PR campaign of words was launched immediately. Claims of a consensus were part of it, but arguments that the debate is over and that the science is settled combined to offset any doubts.

EARLY CONFLICTING EVIDENCE REWRITING HISTORY–THE TROUBLING DIAGRAM

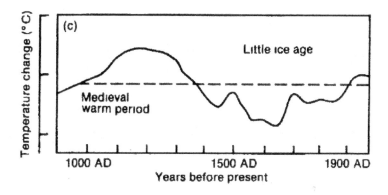

The above graph showing Northern Hemisphere temperatures for the last thousand years appeared as Figure 7c in the first IPCC report in 1990. It posed problems for those who, by 1995, were using climate and especially global warming for their political agenda. It shows how temperature rose to the Medieval Warm Period (MWP) peak around 1180 A.D., then fell to the nadir of the Little Ice Age (LIA) around 1680, and has risen since.

This graph was problematic for the political climate scenario that said the Earth was currently at its warmest because of the human addition of CO2. We will see how they dealt with it later when the so-called "hockey stick" is examined.

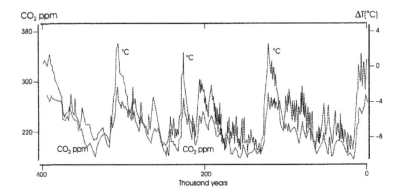

The most important assumption of the AGW hypothesis was that an increase in CO2 would result in a temperature increase. The graph of Antarctic temperature and CO2 derived from air bubbles in the ice spanning 420,000 years was published in 1991. It showed a relationship between CO2 and temperature, and AGW proponents assumed it was presented "proof" that CO2 drove temperature. Just five years later, more detailed analysis showed that temperature change preceded CO2 change.

Everyone now agrees temperature change occurs before CO2 .This juxtaposition is in complete contradiction to the major assumption of the AGW hypothesis and is the relationship in every record of any length, for any period.

The CO2 record does not match the temperature record, and levels much higher than the current 400 ppm existed throughout most of Earth's history.

CO2 AND TEMPERATURE OVER
600 MILLION YEARS

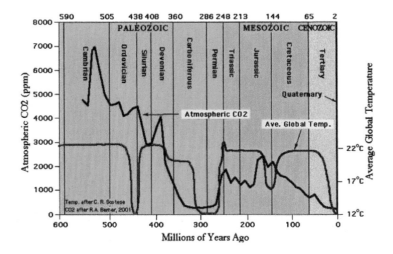

This graph compares temperature and atmospheric CO2, but loses much of the detail of the previous graph. The scale expands dramatically from 420,000 years ago, which is 0.007 percent of the total, to 600 million years ago.

The two curves do not relate at all. Al Gore claimed in his movie that CO2 levels (then at 385 ppm) were the highest ever – in fact, they are the *lowest* in 600 million years. Notice that around 438 million years ago there was an Ice Age (the Ordovician) with CO2 levels at 4000 ppm.

The average level for most of the last 300 million years was 1200 ppm. We know from commercial greenhouses that inject those levels to increase yields by a factor of four, and research by people like Professor Sherwood Idso, that plants thrive best at that level. This suggests plants have evolved to that optimum over time. It is reasonable to argue that plants are malnourished at current levels.

ADJUSTING THE DATA TO
PROVE THE HYPOTHESIS

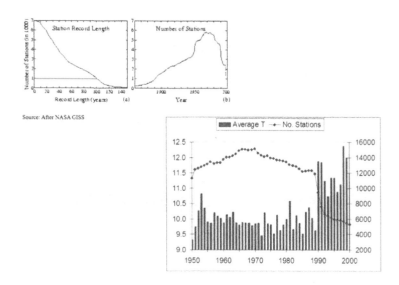

Source: After NASA GISS

The number of weather stations was already inadequate, but was reduced in the 1960s (top right graph (b) under the false assumption that satellites would provide all the data. The World Meteorological Organization (WMO) quickly discovered that many things cannot be measured, such as whether precipitation occurred at all. The top diagram shows the number and length of operation of stations. The red line shows only 1000 weather stations with a record length of 100 years, with almost all in the eastern US or Western Europe.

A significant decline occurred in 1990 when they reduced the number of stations used to calculate monthly and annual global averages. It created an artificial increase in temperatures, as the bottom right graph illustrates. This happened because of fewer stations, but also because of those chosen. The stations are part of the Global Historical Climatology Network (GHCN).

STATIONS CHOSEN TO CREATE THE RECORD

Lack of weather data to build the models is underscored by the further reduction in the number of stations used. Canada, the second-largest country by land mass in the world, illustrates the problem with vast areas with no weather stations.

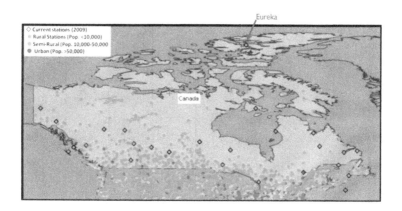

The stations used are in black diamonds. Notice there is only one, Eureka, for the entire subarctic and Arctic, a vast area and the source region for cold arctic air that pushes south and dictates much of the weather across the central and northern half of the continent.

Even more problematic, Eureka is a known anomalously warm spot among all the stations because of local conditions. It does not represent the total region at all.

The lack of stations for adequate temperature coverage is offset by assuming that the weather at any station is representative of a 1200 km (or 745 miles) radius area. Circle any community and see the falsity of this claim. The paucity of coverage is worse for precipitation. For example, attempts to make accurate monsoon forecasts for the Sahel region in Africa failed. The authors concluded:

> *One obvious problem is a lack of data. Africa's network of 1152 weather watch stations, which provide real-time data and supply international climate archives, is just one-eighth the minimum density recommended by the World Meteorological Organization (WMO). Furthermore, the stations that do exist often fail to report.*

DATA MANIPULATION

MAKE THE DATA FIT
ANYHOW

In order to emphasize the claimed impact of human warming, it became standard practice for all national weather agencies to reduce the historic temperature readings. This changed the slope of the temperature curve to show greater warming than was actually occurring. The curve fit and emphasized the political narrative.

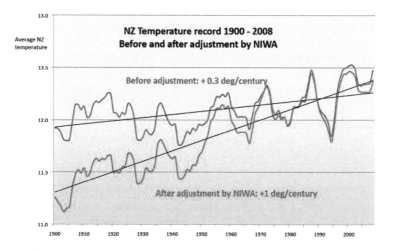

The graphs show adjustments to all New Zealand records, top right. A lawsuit was brought against NIWA, the government agency responsible for these unwarranted adjustments. The government suggested a commission of inquiry and used the Australian Bureau of Meteorology (ABM). The problem is ABM was doing the same thing, as the Rutherglen example illustrates.

Rutherglen, Victoria, Annual Average Minimum Temperatures

Blue is pre homogenization showing cooling of 0.35 degree Celsius per century.
Red is post homogenization showing warming of 1.73 degree Celsius per century.
Clearly the homogenization process changes the temperature trend very dramatically.

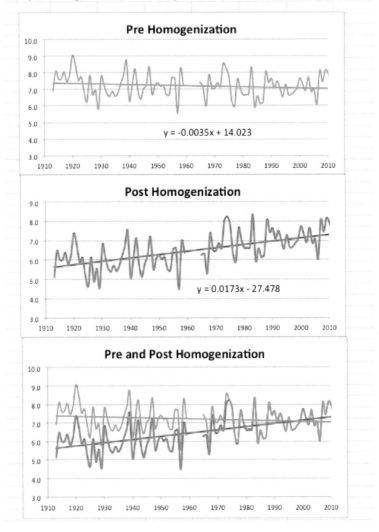

Most adjustments to historic weather records are unnecessary. They are all designed to lower the early temperature readings, thus increasing the amount of warming. The larger problem is the weather records are inadequate even if they are adjusted. *Remember, there are virtually no stations for at least 80 percent of the Earth.* Most stations are concentrated in eastern North America and western Europe, but even those are of less than 100 years old. There are virtually no stations for the 70 percent that is ocean. On the land, there are virtually no stations for the 20 percent that is mountains, the 19 percent grasslands, and the 40 percent forests. The data is inadequate as the basis for building computer models and scientific analysis and definitely must not underpin global changing energy and environment policies.

ADJUSTING THE DATA TO FIT THE HYPOTHESIS

BEST ADJUSTMENTS EVAH

The scientific method involves scientists speculating about how things work. They call the speculation a hypothesis. It is based on a set of assumptions and is only as valid as assumptions. Other scientists, then, performing their proper role as skeptics, try to prove the hypothesis wrong. They do it by challenging the assumptions. Prove them wrong and the hypothesis fails.

The need to prove the AGW hypothesis steered he IPCC on a course of manipulation. The outcome of the IPCC research was predetermined by the definition of climate change, the data selected, and the construction of the computer models. It also involved marginalizing scientists who tried to practice proper science. This meant evidence that contradicted what they wanted had to be altered or counteracted. The following are a few major examples.

CHANGING THE IPCC REPORTS

Full-scale manipulation began when lead author of Chapter 8 of the 1995 report, Benjamin Santer, changed comments previously agreed to by chapter authors:

> *None of the studies cited above has shown clear evidence that we can attribute the observed [climate] changes to the specific cause of increases in greenhouse gases.*

> *While some of the pattern-base discussed here have claimed detection of a significant climate change, no study to date has positively attributed all or part of climate change observed to man-made causes.*

> *Any claims of positive detection and attribution of significant climate change are likely to remain controversial until uncertainties in the total natural variability of the climate system are reduced.*

SANTER'S INSERTIONS

There is evidence of an emerging pattern of climate response to forcing by greenhouse gases and sulfate aerosols ... from the geographical, seasonal and vertical patterns of temperature change ... These results point toward a human influence on global climate.

The body of statistical evidence in chapter 8, when examined in the context of our physical understanding of the climate system, now points to a discernible human influence on the global climate.

He cherry picked the graph as shown to make his claim.

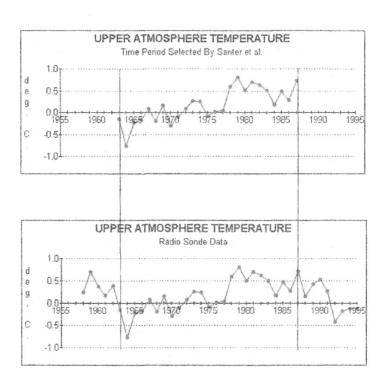

Avery and Singer noted in 2006:

> *Santer single-handedly reversed the 'climate science' of the whole IPCC report and with it the global warming political process! The 'discernible human influence' supposedly revealed by the IPCC has been cited thousands of times since in media around the world, and has been the 'stopper' in millions of debates among nonscientists.*

TROUBLESOME GRAPHS

This 7c graph, produced by Hubert Lamb that appeared in the 1990 IPCC report, was a problem requiring correction because it showed it was warmer 1000 years ago during the Medieval Warm Period than today. In testimony to the US Congress, Professor David Deming provided amazing testimony:

> *With the publication of the article in Science [in 1995], I gained significant credibility in the community of scientists working on climate change. They thought I was one of them, someone who would pervert science in the service of social and political causes. So one of them let his guard down. A major person working in the area of climate change and global warming sent me an astonishing email that said,*
>
> *'We have to get rid of the Medieval Warm Period.'*

The reported source of the email was a senior member of the Climate Research Unit (CRU) and the IPCC.

This means they planned to deliberately rewrite history. It became the centerpiece of the 2001 IPCC report, despite warnings from an independent report prepared by Professor Wegman for a congressional inquiry.

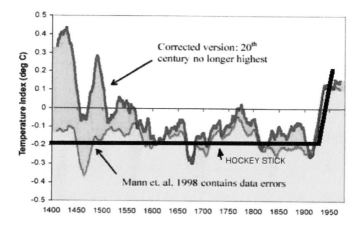

The red line indicates the graph that was created and published in the 2001 IPCC report. The three author's labeled it "the hockey stick" because of the general shape (superimposed in green). The 'stick' eliminated the MWP and the Little Ice Age (LIA). Two Canadian researchers, Steve McIntyre and Ross McKitrick, pursuing the standard scientific method, tried to reproduce results shown with the blue line. They recognized what was manipulated and with limited data were able to reconstruct what was done by Michael Mann, Bradley, and Hughes.

The 'stick' was created with deliberately selected tree rings, which represent precipitation, not temperature, and when they showed a downturn in the 20th century, they threw out the data and patched on 20th century thermometer data. The perpetrators referred to this process as "hide the decline" and the patching of data as "Mike's nature trick", after the first publication in which it appeared. Mann has consistently refused to disclose his original data and calculations. Author and journalist Mark Steyn wrote a book about Mann's behaviour titled, "A disgrace to the profession".

THE ACTUAL DIAGRAM

The original graph illustrates more clearly what was done to hide the decline. The downward trend from 1000 to 1900 continued into 2000, so they tacked on the blade, which is comprised of instrumental temperature data. The splicing alone is unscientific, but the temperature data itself is very questionable. It was produced by Phil Jones, Director of the Climatic Research Unit (CRU and a key member of the IPCC).

He claimed that the global average surface temperature increased by 0.6°C since the late 19th century. The earlier graph about the number of stations showed there are fewer than 1000 stations, over 100 years old, and they are concentrated in eastern North America and Western Europe. In addition, they are all recording land temperatures, which are always higher than ocean temperatures. Jones acknowledged the limitations of the data, but everybody ignored it. The actual data was a 0.6°C increase, but the error range was ±0.2°C or ±33%, which means they are of no scientific value.

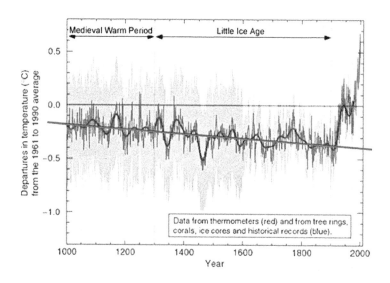

When asked by Warwick Hughes to disclose his data, Jones wrote on 21 February 2005,

> *We have 25 or so years invested in the work. Why should I make the data available to you, when your aim is to try and find something wrong with it?*

We will examine later how Jones and the CRU gang were exposed for creating scientific evidence using their positions at the IPCC to push the political agenda. It was exactly "the science" Maurice Strong wanted when he created the UNEP and the IPCC.

OMISSIONS: IT'S THE SUN, STUPID

It wasn't just what they put in their science that created the deception; it was what they left out. These omissions are proof of premeditation. They knew what they were doing, which in criminal justice makes the crime more severe.

As noted already, they began with the definition of climate change. They examined only human causes of climate change, which allowed them to leave out major natural mechanisms. They created a system that narrowed the science, amplified the message and, thereby, the threat.

The IPCC was made up of weather bureaucrats. They were assigned to three main groups: Working Group I (WGI) examined the Physical Science Basis; Working Group II (WGII) Impacts, Adaptation and Vulnerability; and Working Group III (WGIII) Mitigation of Climate Change. WGI proves that human CO2 is causing global warming. WGII and III accept that without question and produce the alarmism of the effects, and push a single solution, eliminate fossil fuels.

To reinforce the issue they then produced a Summary for Policymakers (SPM), with a different group of bureaucrats and politicians directed by a few select scientists. The SPM bears no resemblance to the Physical Science report and was released first because they knew few will have read or understood the Science Report. The Santer Chapter 8 example referenced earlier was the first detected example of this type of deliberate difference.

THE MILANKOVITCH EFFECT

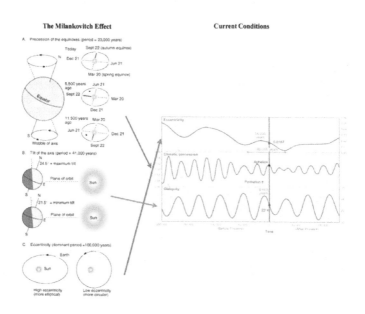

IPCC computer models do not include what is called the Milankovitch Effect, named after the 19th century Serbian mathematician Milutin Milankovitch. These are variations in the three major sun-earth relationships that change the net amount of energy reaching the Earth every single year (right diagram), which changes global temperature. We have known about these changes for over 130 years, but only recently were they introduced in some schools. IPCC computer modeler, Andrew Weaver, told me they're not included because they operate on too long a time scale, but the changes are measurable and significant for their 50 and 100-year predictions. It may seem small, but it is significant, relative to the amount of change human CO2 is claimed to make.

In the diagram, the red arrow connects orbital changes caused by the gravitational pull of Jupiter to the changing eccentricity. The green arrow connects change in tilt and the blue arrow changes in precession or the day on which equinoxes and solstices occur. Change in the net radiation from the Sun is approximately 100 watts per square meter over time, which far exceeds the possible effect of humans.

THE COSMIC THEORY

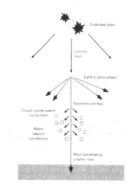

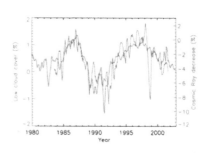

We have known for a long time that global temperatures vary with the number of sunspots. The problem was we didn't have a mechanism. In science, you cannot assume a cause/effect relationship because there is a correlation.

This changed beginning in 1991 when the Cosmic Theory, also known as the Svensmark Theory. It was proposed primarily by Henrik Svensmark, leader of the Center for Sun-Climate Research at the Danish National Space center and explained in *The Chilling Stars; A New Theory of Climate Change*, that he co-authored with Nigel Calder and published in 2007. The theory has since been proved. It shows that sunspot changes are visible evidence of changes in the Sun's magnetic field. As it varies, it controls the amount of cosmic radiation reaching the Earth. When these rays enter the Earth's atmosphere, they become condensation nuclei around which water droplets form. These droplets are what you see collectively as clouds. The graph on the right shows the relationship between low cloud and cosmic radiation. It is the cloud that causes temperature variation.

None of this is included in the IPCC reports or their computer models.

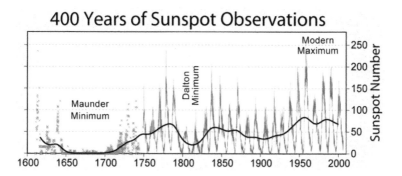

This diagram shows the plot of sunspots and their variation from 1610 forward (Galileo's first observations with the telescope). Currently, sunspot numbers are declining as Cycles 23 and 24 show.

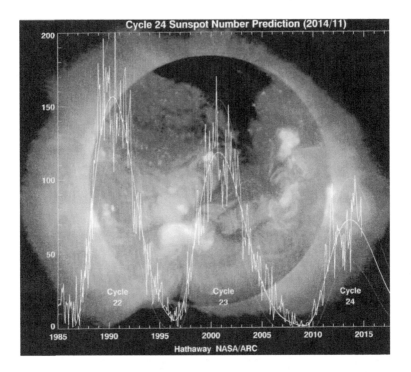

This is what is causing the current cooling trend. Indications are we will reach temperatures similar to those associated with the Dalton Minimum (1790 - 1830), but many believe it will be similar to the Little Ice Age associated with the Maunder Minimum (1640 - 1710).

HISTORIC EVIDENCE

The painting *Thames Frost Fair* by Jan Grifier shows the River Thames in 1683 with 1 meter (3 feet) of ice. It shows a medieval "Ice Fayre" that were popular during the reign of Elizabeth I. The last fair occurred around the 1830s as the Little Ice Age was changing to the modern warm period.

The world has not experienced any warming for 1998 to the present. Claims that the warmest years are still occurring are a statistical deflection. Of course, the warmest years are going to occur at the top of a curve, even though a declining trend has begun. Meanwhile, severe weather has not increased, but temperatures show greater fluctuations, which are also a natural phenomenon as the upper atmosphere planetary air flow patterns change.

THE CIRCUMPOLAR VORTEX

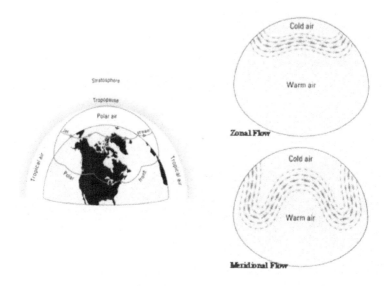

AGW proponents present normal weather events as abnormal. Current patterns of extremes of temperature, with cold air pushing toward the equator, are normal.

The atmosphere is comprised of two air masses, cold polar air and warm tropical air (left diagram). They meet along the Polar Front. Very strong winds blow along the Front in the upper atmosphere, known as the Circumpolar Vortex and included Rossby Waves, named after meteorologist Carl Rossby, who first identified them in 1946.

As the world cools down, the Polar Front moves toward the Equator and the wave pattern switches from Zonal to Meridional (right diagram). When the waves become deeply meridional, they tend to block, that is their normal movement from west to east stalls, causing prolonged spells of hot, cold, wet or dry weather.

John Holdren, Obama's Director of Science and Technology, made a two-minute video implying that what he incorrectly called the Polar Vortex explained the cold outbreaks and was due to warming.

It is remarkable that a person in Holdren's position could present such false information as fact, but it is the treadmill you are on when you continue to try and prove rather than disprove a scientific hypothesis.

SOCIAL AND ECONOMIC DAMAGE

CO_2 is not causing global warming or climate change. A cabal is an obscure political clique or faction that promotes their agenda. It was a cabal, known as the Club of Rome (COR) and formed in 1968, that promulgated the global warming agenda. People involved with the COR, and marshalled by Maurice Strong, contradicted the fact about CO_2 for their political agenda to claim that it was causing global warming, and so deliberately created the myth.

The Club of Rome is a Neo-Malthusian organization with interlocking membership with European power elite groups such as the Committee of 300 (a secret society founded by the British aristocracy in 1727) and the Bilderberg Group.

They achieved their goal, and so politicians and governments wanting to appear "green" established policies to eliminate or at least dramatically reduce fossil fuels. The deception is encapsulated in the misuse of the term greenhouse gases when discussing CO_2 as having a negative impact on the welfare of the Earth. In fact, governments have only focused on CO_2 while ignoring H_2O, the largest and most important greenhouse gas. The problem is that the term "greenhouse gas" has a negative connotation. The entire exercise became became known as the Green Agenda and has failed everywhere it has been tried from Spain to Ontario, Canada.

This book identifies key people who developed and activated the anti-overpopulation, anti-development, anti-free enterprise, and pro-one world government policies of the Club of Rome. Maurice Strong was critical to advancing the idea that was CO_2 causing global warming and destroying the planet because he was the conduit who made it happen through the UN. In 1992, the same year he chaired the UNEP conference in Rio de Janeiro, Maurice Strong became Chair of Ontario Hydro, the public utility that controls all energy production for the province.

Ontario has been what Canadians call a "have" province. It has been the powerhouse economy since the formation of Canada in 1867. Single-handedly, Strong took care of the economic strength by putting in place the Green Agenda he planned for the world. He cancelled planned nuclear power plants and a scheduled phase out for those in operation. He reduced and closed coal fuelled stations and cancelled plans for future plants. He built wind turbines that are unable to produce the power required and which need 100 percent fossil fuel back-up in case the wind stops blowing. He made bad and potentially ruinous financial investments with money. Ontarians pay a premium on their monthly energy bills and will pay for decades to come. They experience and suffer just as any country that has tried the green agenda has done. All those countries, such as Spain, Germany, and Britain are abandoning the agenda, just as many who agreed to the agenda at the 2015 Paris Climate Conference forego their commitments.

WHO?

After the publication of the first complete image of Earth taken by astronauts on Apollo 8 appeared, momentum for the new paradigm of environmentalism accelerated. The image emphasized the small finite nature of the Earth, and names like Spaceship Earth or the Blue Marble appeared on posters, book covers and in school textbooks undermining the tenuous nature of Earth and our existence as passengers. The problem was that a few extremists with political ambition grabbed the notion for political control. They used it as a vehicle to take the moral high ground, to claim that only they cared about the environment. They argued that everyone else was guilty of environmental destruction because of their avarice and wasteful ways.

Contrary to what most people think environmentalism, as a political agenda, it is pushed by extremely wealthy and powerful left wing people who make their money exploiting the environment. The psychology of that is beyond the discussion here, but consider the hypocrisy of George Soros, Maurice Strong, the Rockefellers, Leonardo DiCaprio, and Ted Turner, among many others. George Soros is a multi-billionaire who funds groups like the Tides Foundation that undermine economies then he profits from the collapse. DiCaprio owns five large houses and uses private jets to go round the around the world urging constraint and restrictions in the lives of ordinary hard-working people. Ted Turner lives a similar profligate lifestyle while he hired former US Senator Timothy Wirth to a high paid position at the UN. While still in the Senate Wirth said, "We've got to ride the global warming issue. Even if the theory of global warming is wrong, we will be doing the right thing ..." The confusion is similar to that about another person, who most people think was a right wing fascist, but was a socialist promoting environmentalism, Adolf Hitler. Nazi stands for National Socialism and as one author explained:

> *It is a well documented though seldom highlighted fact that the Nazis were very much into environmentalism— not for environmentalism's sake, of course, but rather as a means of oppression and control. As it turns out, environmentalism fits the form of tyranny like a well tailored suit.*

The wealthy and powerful people speak and act individually, but collectively many of them acted through their privileged group called the Club of Rome (COR). The founding meeting of the Club occurred in 1968 at David Rockefeller's estate in Bellagio, Italy.

Although the process to use environmentalism and global warming for a political agenda began earlier, the objective was set out by Alexander King, an eminent Scottish scientist and President of the COR, who joined with his Assistant Secretary, Bertrand Schneider, in the 1991 publication *The First Global Revolution*, which said:

> *In searching for a common enemy against whom we can unite, we came up with the idea that pollution, the threat of global warming, water shortages, famine, and the like,*

would fit the bill. In their totality and their interaction these phenomena do constitute a common threat which must be confronted by everyone together. But in designating these dangers as the enemy, we fall into the trap, which we have already warned readers about, namely mistaking symptoms for causes. All these dangers are caused by human intervention in natural processes, and it is only through changed attitudes and behaviour that they can be overcome. The real enemy then is humanity itself.

The list of enemies or threats to the planet was designed to unite people. In reality, the challenge was to overcome individual nation-states that might oppose the establishment of one-world government or global socialism. The idea is that global warming transcends national boundaries making it a global problem that national governments cannot resolve. The changed behaviour they sought was for everyone to become a socialist under a one-world government.

The authors claim that the list of enemies is designed to unite people. In fact, it is needed to overcome what they see as the divisiveness of nation-states and to justify the establishment of a one-world government. They claim that global warming is a global problem that national governments cannot resolve.

People find it difficult to believe that only a few people could or would deceive the world. They shouldn't because as social anthropologist Margaret Mead warned, "Never doubt that a small group of thoughtful, committed citizens can change the world. Indeed, it is the only thing that ever has."

People tend to accept the adage that no one person can change the world. They are more likely to hold that view in the internet age as conspiracy theories abound. Generally speaking, there are also groups in society who are thought to be above mundane human weaknesses of succumbing to greed and power. Scientists are one of these groups; clearly that was Klaus-Eckart Puls' thinking when he accepted the IPCC science without question. Fortunately, this is rapidly changing as more scientific corruption and incompetence is exposed every day. Why would scientists pervert and corrupt science? For the same reasons as anyone else, personal gain that is either financial, political, or both.

What the scientists did wasn't illegal; everybody is entitled to their political views and strategies. The illegality and immorality are in how they manufactured a crisis and used power, influence, and deception to achieve their goal.

There is a big stretch between formulating an objective and putting it into effect. The ideal person was Maurice Strong, an active member of the COR. In 2001, Neil Hrab, a long time observer of Strong, explained his talents of organizing, manipulating and "...mainly using his prodigious skills as a networker. Over a lifetime of mixing private sector career success with stints in government and international groups..."

MAURICE STRONG

In 1966, Strong became head of the Canadian International Development Agency (CIDA) and thus he was launched internationally. Impressed by his work at CIDA, UN Secretary General U Thant asked him to organize what became the first Earth Summit – the Stockholm Conference on the Human Environment in 1972. The next year, Strong became first director of the new UN Environment Program (UNEP), created as a result of Stockholm.

In a prophetic speculation, Strong was cited in Elaine Dewar's book *Cloak of Green* saying, "Isn't the only hope for the planet that the industrialized civilizations collapse? Isn't it our responsibility to bring that about?"

After spending several days with Strong at the UN, Dewar concluded that, "Strong was using the UN as a platform to sell a global environment crisis and the Global Governance Agenda."

In 1992 Strong became Chair of Ontario Hydro, the government agency that controls all energy in the Canadian province. His position there is an insight into what happens when the official policies expressed by the UN in the 2015 Paris Climate Conference are applied. He applied his UN style energy policies and drove Ontario from one of the most powerful and wealthy provinces in Canada to one of the weakest because of a lack of power resources, growing manufacturing and living costs, and increasing debt. This is not unique; any country, such as Spain, Denmark, or Germany that adopts what is collectively called a green agenda, quickly runs into trouble.

JOHN HOLDREN

John Holdren is Obama's Head of Office of Science and Technology Policy. His academic career focused on the central themes behind COR thinking that the world is overpopulated and we are quickly exhausting all the resources. His career in the White House pursues those themes but under the guise of global warming.

The worst offenders of resource exploitation are those in the industrialized developed nations who are doing it at an unsustainable rate using fossil fuels, hence the focus on them and their byproduct CO2. These ideas and possible solutions are encapsulated and linked to the environment in the 1977 book *Ecoscience: Population Resources,*

Environment that John Holdren co-authored with Paul Ehrlich. They wrote:

- *Women could be forced to abort their pregnancies, whether they wanted to or not.*

- *The population at large could be sterilized by infertility drugs intentionally put into the nation's drinking water or in food.*

- *Single mothers and teen mothers should have their babies seized from them against their will and given away to other couples to raise.*

- *People who "contribute to social deterioration" (i.e. undesirables)... "can be required by law to exercise reproductive responsibility" -- in other words, be compelled to have abortions or be sterilized.*

- *A transnational "Planetary Regime" should assume control of the global economy and also dictate the most intimate details of Americans' lives -- using an armed international police force.*

In the book, Holdren discloses the technique of making a truly frightening and draconian policy palatable, necessary and workable. He explains how to bypass US legislation designed to prevent imposition of such freedom destroying totalitarian ideas. He wrote:

> *Indeed, it has been concluded that compulsory population-control laws, even including laws requiring compulsory abortion, could be sustained under the existing Constitution if the population crisis became sufficiently severe to endanger the society.*

The question is who *"concluded that compulsory population-control"* could be sustained? The answer is the authors did. The next question is who decides *"if the population crisis became sufficiently severe to endanger the society"* ? Again, the authors. So, they claim there is a problem, then they decide when it is severe enough to warrant complete suspension of legal controls to justify totalitarian actions.

At his 2009 Senate confirmation hearing, Holdren claimed to renounce these views. However, his actions don't support that claim because he is using climate and environment from his position in the White House to pursue the same objective. He is the source of such public relations terms like "Climate Disruption" and the scientifically incorrect "Polar Vortex." Although used in some ill-informed sources, such as *Scientific American,* which consistently presents incorrect sensational information in their push to sell science on the newsstands, Polar Vortex is properly called the Circumpolar Vortex. Holdren, and now others, use it to refer to one outbreak of cold polar air, when that is actually just one Rossby Wave along the Circumpolar Vortex. This is part of the public relations game of selling false information such as slogans or names like "global warming skeptic" and "climate change denier", with all the Holocaust connotations to direct thinking. The objective is to isolate an idea or individual, thus implying those isolated are not in line with normal thinking. If a specific term, such as "birther" for anyone who dares to ask about President Obama's resume, is not created, then the general category of conspiracy theory is used.

AL GORE

Nobody achieved a higher profile on the global warming issue than Al Gore. Like anyone with political ambition, he sought a vehicle to advance his political and financial fortunes. He chose the environment, generally, and global warming, specifically. As he said:

> *I vowed always to put my family first, and I also vowed to make the climate crisis the top priority of my professional life.*

In a bizarre New York Times Op-ed titled *"We can't wish away climate change"(2010)*, Gore included all the standard errors that entrap and confuse most people. The title illustrates how little Gore knows or understands. No, we can't wish it away because climate change has and will always exist. It is pointless to elaborate on Gore's involvement and exploitation of the climate issue. Just two points summarize what is wrong with everything he says and does regarding climate. The first is that he shared a Nobel Prize with the IPCC, but they disagreed significantly on sea level rise. Gore claims he is right and the IPCC, wrong:

> *It is worth noting that the panel's scientists – acting in good faith on the best information then available to them*

– probably underestimated the range of sea-level rise in this century.

They are both wrong, but Gore by a much greater margin. Sea level is essential to Gore's misdirection and fear factor. Ask people what are the negative effects of global warming and they inevitably say sea-level rise. It is a major part of Gore's 2006 propagandistic movie *An Inconvenient Truth* showing computer reconstructions of water flooding over Florida and other regions.

The second happened in the same week he received the Nobel Prize, a UK court uudge, Justice Burton, released a ruling on Al Gore's documentary. The case was brought forth by a father against the UK Department of Education. The following quotes are from the official 2007 UK court ruling:

> *Stuart Dimmock is a father of two sons at a state school and a school governor. He has brought an application to declare unlawful a decision by the then Secretary of State for Education and Skills to distribute to every state secondary school in the United Kingdom a copy of former US Vice-President Al Gore's film.*

The Judge's ruling said:

> *I viewed the film at the parties' request. Although I can only express an opinion as a viewer rather than as a judge, it is plainly, as witnessed by the fact that it received an Oscar this year for best documentary film, a powerful, dramatically presented and highly professionally produced film. It is built round the charismatic presence of the ex-Vice-President, Al Gore, whose crusade it now is to persuade the world of the dangers of climate change caused by global warming. It is now common ground*

that it is not simply a science film – although it is clear that it is based substantially on scientific research and opinion – but that it is a political film, albeit, of course, not party political. Its theme is not merely the fact that there is global warming, and that there is a powerful case that such global warming is caused by man, but that urgent, and if necessary expensive and inconvenient, steps must be taken to counter it, many of which are spelt out.

The ruling also identified nine scientific errors in the film, none of which Gore has corrected.

WHAT?

UNDERLYING MOTIVATION: OVER-POPULATION

The focus of the debate is on global warming and CO2, that the latter causes the former, but that is a major part of the deception. The real target is what some claim is an overpopulated, industrialized, exploitive world using too many resources too quickly and unsustainably. In other words, too many people equals too much CO2.

The notion is a development of the Thomas Malthus idea that the world population would outgrow the food supply explained in his 1796 book *An Essay on the Principle of Population*. Charles Darwin took a copy with him on the *Beagle for* the voyage on which he gathered evidence and formulated the idea of evolutionary theory. Paul Johnson, author of a Darwin biography, described Malthus's work as *"Darwin's emotional detonator"*. Malthus' book influenced Darwin enormously, but unlike his own demand for rigor in science, he completely overlooked its lack in Malthus'. The examples Malthus used, such as population growth in the United States, were totally inappropriate and wrong. This theme of overpopulation is still driving the entire environmental movement, especially the global warming threat, but it is still completely wrong, which I'll point out shortly.

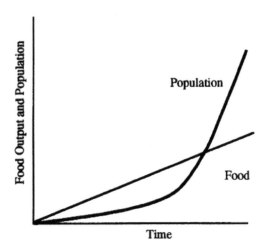

Malthus believed that human behaviour was influenced by govern-ment policies that encouraged them to breed and were thus a negative thing. As Johnson explains, Malthus's aim was to discourage charity and reform the existing "poor laws" policies. He argued that the laws encouraged the destitute to breed, and so aggravated the problem of overpopulating with low quality people going forward. The ideas are redolent with eugenics, an idea and word initiated by Darwin's half-cousin, Francis Galton. Encyclopedia Brittanica defines eugenics as "the selection of desire heritable characteristics in order to improve future generations, typically in reference to humans."

As discussed earlier, the work of John Holdren and Paul Ehrlich was about eugenics and the legislative creation of fewer more desirable people. Their work was central to and influenced the Club of Rome's concern with overpopulation. The Club extended and expanded these ideas amongst very powerful people:

The links between powerful people and control of the population reso-nated throughout the Enlightenment. This is the period that created the scientific revolution, and today's world of individual freedom along with Emmanuel Kant's comments "Dare to know! Have the courage to use your own reason." The paradox is the conflict between wealth that emerges from capitalism and total control of the people, which is communism. Under the latter you don't question, but now the wealthy, such as those listed earlier, don't want questions either.

The first Heartland Conference Climate Conference in New York in 2004 provided the first international forum for those scientists performing their proper function as skeptics to challenge the government claims. Vaclav Klaus, Prime Minister of the Czech Republic, was the keynote speaker. His opening remark was that the Czech people have just gone through 70 years of communism, so why the hell would they want to go back to that? His presentation explained how, from that experience, he saw environmentalism and global warming being used for total control. He summarized these views in a 2008 book titled, *Blue Planet in Green Shackles*. It supports the fact that environmentalism and AGW share a political agenda pushed by extremely wealthy and powerful left wing people, most of whom made their money exploiting the environment. In a 1976 interview with Maclean's magazine, Maurice Strong described himself as 'a socialist in ideology, a capitalist in methodology'.

This description appears to apply to all the wealthy people listed earlier, like George Soros and Bill Gates. It also applies to entire countries. For example, China and Russia practice "state capitalism" with centralized governments controlling a capitalist economy.

OVERPOPULATION CLAIMS IN THE 20TH CENTURY

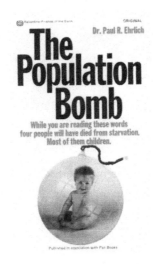

Paul Ehrlich made overpopulation a global issue and concern with his 1968 book, *The Population Bomb*. The book title exploits emotionalism. "While you are reading these words four people will have died from starvation. Most of them children." This statement is pure speculation because data did not exist to make such a claim. Overpopulation wasn't a problem. Ehrlich admits humans occupy no more than 3 percent of the earth's land surface. Success of his duplicity is that few people believe that statistic. The reality was that the activities of some people offended his political views. On April 6, 1990, the Associated Press quoted him as follows, "Actually, the problem in the world is there is much (sic) too many rich people."

Dixy Lee Ray's 2011 book *Trashing the Planet* quotes Ehrlich as follows, "We've already had too much economic growth in the United States. Economic growth in rich countries like ours is the disease not the cure."

That is untrue as the economic model known as the *Demographic Transition* indicates. In every developed country a sequence of a death rate decline is followed by a birth rate decline, culminating in a net population decline.

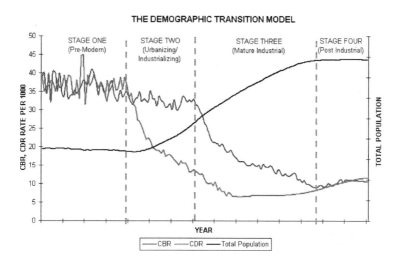

Paul Ehrlich's ideas and credibility are totally discredited by the failure of his dire predictions. A theme we will see throughout the warming deception story is that a simple definition of science is the ability to predict. If the prediction is wrong, the science is wrong. For example, he wrote:

The battle to feed humanity is over. In the 1970s, the world will undergo famines. Hundreds of millions of people are going to starve to death in spite of any crash programs embarked upon now. Population control is the only answer.

By 1994, despite Ehrlich's failed predictions, the United Nations held a world population conference in Cairo, Egypt. This point is important throughout this story because if your predictions are wrong your science or assumptions and models are wrong. They should not become the basis of any policy. But the conference was, as with all these alarmist claims, about the politics not the facts. Chief political exploiter and cheerleader for the Conference was Al Gore, keynote speaker and leader of the US delegation. Cairo was the ideal location with the hungry millions outside the conference hall. Participants ignored the fact that the Netherlands has one of the the highest population densities and highest standard of living in the world.

THE WORLD IS NOT OVERPOPULATED!

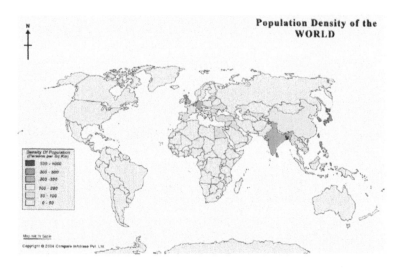

Central to the Club of Rome neo-Malthusian theme and justification for their actions is that an overpopulated world is exhausting the resources. It was encapsulated in their 1972 publication, *Limits to Growth*. It is also reflected in the term "peak oil", the idea that we are close to the point where we will increasingly run out. Now we know that it is incorrect, that it was myth created to perpetuate the political agenda. It is like all the myths created to extend and amplify the narrative that there too many people using too much. Perhaps the biggest myth is that the world is overpopulated.

Most of the world is unoccupied with most people concentrated in flood plains, such as in China along the Yangtze (Chang Jiang) River and deltas, like the Nile in Egypt. Canada is the second largest country in the world with approximately 33.6 million residents (2009). California had a 2008 population of 36.8 million people. Statistically, the entire population fits certain islands, such as the Isle of Wight off the south coast of England or regions, like half of Alberta. For example, Texas at 7,438,152,268,800 square feet divided by the 2015 world population of 6,774,436,692 gives 1098 sq. ft. per person. Fitting them in is different from the ability to live there. Population geographers distinguish between ecumene, the inhabited area, and non-ecumene, the uninhabited areas. Habitable areas change all the time. The area of the earth that is habitable has changed because of technology, communications, and food production capacity.

The Malthusian notion of too many humans for too little food galvanized the modern debate. It is a false issue used by a left wing ideologists that needed an overarching global issue. As stated in the Club of Rome: The First Global Revolution (1991):

> *The common enemy of humanity is man. In searching for a new enemy to unite us, we came up with the idea that pollution, the threat of global warming, water shortages, famine and the like would fit the bill. All these dangers are caused by human intervention, and it is only through changed attitudes and behavior that they can be overcome. The real enemy then, is humanity itself.*

The Club of Rome expanded the Malthusian theme from running out of food, to running out of all resources. It was encapsulated in a book titled *Limits to Growth* (1972).

The threat had to be global to transcend national boundaries and thus argue for a single global government as the only option to save the planet.

The Club of Rome needed a focus and numerical evidence that appeared scientific. They could then use this to justify political action. This is normal practice except that it is properly based on facts and validated science. In this case they created the facts they needed, thus creating a myth to justify. In his 2008 book *True Enough: Learning to live in a Post-Fact Society,* Farhad Manjoo claims, "Facts no longer matter. We simply decide how we want to see the world and then go out and find experts and evidence to back our beliefs." It wasn't difficult to find support because as Niccolo Machiavelli said, "One who deceives will always find those who allow themselves to be deceived."

HOW

ACHIEVING THE GOAL

This is the largest section of the book because the major weapon in the deception that global warming is due to human production of CO2 involved the creation of the political and scientific structure to predetermine the outcome. The people organizing the deception knew few would understand and they could easily marginalize those who did.

In her book *Cloak of Green*, Elaine Dewar asked Maurice Strong why he didn't enter politics to achieve his goal of causing the collapse of the industrialized nations. He said you can't do anything as a politician. Strong said he was going to the UN where, "He could raise his own money from whomever he liked, appoint anyone he wanted, control the agenda."

Dewar's book is a classic example of good investigative journalism. It also shows that environmentalists are not what they claim. They use the environment as a vehicle to push a socialist, total government controlled, agenda. Dewar planned a book praising Canadian environmentalists including Elizabeth May, David Suzuki, and Maurice Strong. Dewar kept an open mind and did not set out to prove her hypothesis. The research showed that these Canadian environmentalists were more corrupt and manipulative than the people they attacked. May and Suzuki misused and misrepresented data ostensibly for saving the planet but in reality for a political agenda

The bigger question is how to bring about the collapse of industrialized nations. Strong knew exactly what to do and how to do it as I will explain later. He also knew the key was controlling the bureaucracy, as Henry Lamb explained in a 1997 said:

> Strong prefers to operate in the background. He, perhaps more than any other single person, is responsible for the development of a global agenda now being implemented throughout the world.

He concluded:

> The fox has been given the assignment, and all the tools necessary, to repair the henhouse to his liking.

This is why Dewar concluded, after spending a few days with Strong at the UN that, "Strong was using the UN as a platform to sell a global environment crisis and the Global Governance Agenda."

BUREAUCRACY

"Bureaucracy, the rule of no one, has become the modern form of despotism."

Mary McCarthy, Author, Critic and Political Activist.

"It's always cozy in here. We're insulated by layers of bureaucracy."

Strong was a senior member of the Club of Rome with the skills to translate their ideas to political reality. He combined them with his organization and networking skills with powerful people and translated them into United Nations legislation.

He made them the basis of the United Nations Environmental Program (UNEP). Agenda 21 became the political portion and the Intergovernmental Panel on Climate Change (IPCC) was the scientific vehicle. He introduced them through the 1992 Rio Conference, which encompassed the public, media, and the environmental movement through the Non-Government Organizations (NGOs). This was important politically, but the most important step for control was involving the World Meteorological Organization (WMO), the body made up of weather office bureaucrats from every UN member nation. But, more of that shortly.

Now the global stage was set to save the planet from the demons of industry. The mob was aroused, and the essential slogans were created. Strong promoted some of these, such as Rene Dubos' *"Think globally, act locally."* or *"Sustainable Development"* from Gro Harlem Brundlandt's report, *Our Common Future.*

ORGANIZATIONAL CHART: A BUREAUCRAT'S DELIGHT

At the Rio 1992 conference he put in place the two arms necessary to create and control global warming as a political agenda (Agenda 21 provides the political and philosophical basis. The United Nations Framework Convention on Climate Change (UNFCCC) set the ground rules for climate science to pre-determine their claim that CO2 was the problem. The Conference of the Parties (COP) make the political and economic decisions based on the science the IPCC creates for them. The first COP (#1) was held in Berlin in 1995. The most recent was COP #21 in Paris in November 2015 at which 195 nations appeared to solve the problem of climate change.

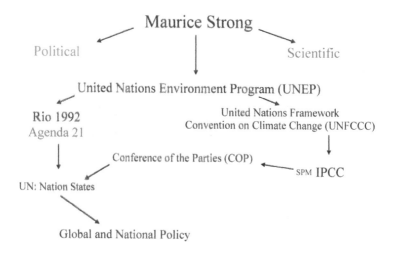

HOW POLITICIANS AND BUREAUCRATS DEFLECT, BUT RETAIN CONTROL

I used to think, like most, that Commissions of Inquiry were wonderful vehicles that removed politics from a dispute and sought the truth. They vary in form and format but generally are created by national governments. In Canada for example, "Commissions of Inquiry are established by the Governor in Council (Cabinet) to fully and impartially investigate issues of national importance." Regardless of jurisdiction, Commissions of Inquiry are all used to control, although most people think they are independent and at arms length from politics. When a politician appoints a commission everybody thinks it removes the politics and political control. Politicians are off the hook because they say they cannot comment on an issue until the Commission reports on it. They go back to their offices and with the bureaucrats define terms of reference that predetermine the outcome.

I know this because it happened to me on the very first of many Commissions regarding conflict over water use and levels in a lake. The terms of reference were so restrictive that the Commission didn't have access to the basic historical data. I told the chairman of the Commission to tell the Minister of Natural Resources that unless we received unfettered access to all information, I would tell the media he was trying to predetermine the outcome. I got the information and discovered that the problems on the lake were identified and the solutions provided 100 years earlier, and by two intervening commissions. The politicians didn't care; they deflected the problem from troubling their careers and were able to avoid action by pushing the issue "down the road", meaning to their political descendants.

Earlier, I identified what Maurice Strong said was the global problem. "Isn't the only hope for the planet that the industrialized civilizations collapse? Isn't it our responsibility to bring that about?"

The question is how do you bring that about? The answer is understood by comparing the industrial nation to a car engine. They both run on fossil fuels. You can stop the engine by stopping the fuel supply. However, if you do that the people will react quickly and negatively – it is politically dangerous. However, you can also stop the engine by plugging the exhaust. Carbon dioxide (CO_2) is a major release from fossil fuels. If you show that it is causing global warming, that is destroying the planet, you justify stopping its production and use. This would shut down the industrialized nation.

Maurice Strong and his cabal narrowed the focus and research of the Intergovernmental Panel on Climate Change (IPCC) by the definition of climate change in Article 1 of the United Nations Framework Convention on Climate Change (UNFCCC).

A change of climate which is attributed directly or indirectly to human activity that alters the composition of the global atmosphere and which is in addition to natural climate variability observed over considerable time periods.

The problem is you cannot determine human effect unless you know the amount and cause of natural climate change, and we don't.

The UNEP predetermined the outcome of the IPCC by limiting their definition of climate change to only human causes. They also created the illusion that they were the "official" word on all climate matters.

STRUCTURE OF THE IPCC

Strong set up the IPCC through the World Meteorological Organization (WMO) as a masterstroke essential to controlling world governments on climate issues. The WMO is made up of weather office bureaucrats from all UN nations. For example, a senior bureaucrat from Environment Canada chaired the 1985 meeting in Villach, Austria to establish the IPCC. These bureaucrats pushed the IPCC agenda on their politicians. It worked because they knew politicians didn't understand the science. Those who dared to question were easily confronted. MIT Professor Richard Lindzen said:

> *It is no small matter that routine weather service functionaries from New Zealand to Tanzania are referred to as "the world's leading climate scientists." It should come as no surprise that they will be determinedly supportive of the process. IPCC's emphasis, however, isn't on getting qualified scientists, but on getting representatives from over 100 countries, said Lindzen. The truth is only a handful of countries do quality climate research. Most of the so-called experts served merely to pad the numbers.*

Scientists who questioned were marginalized by the claim that a consensus has nothing to do with science. They could make incorrect statements and marginalize anyone who challenged their consensus. Equally important, the bureaucrats who represented their nations as members of the IPCC directed national funding to people who supported the scientific and thereby the political agenda. As Upton Sinclair said, "It is difficult to get a man to understand something, when his salary depends upon his not understanding it."

HOW SO FEW FOOLED SO MANY

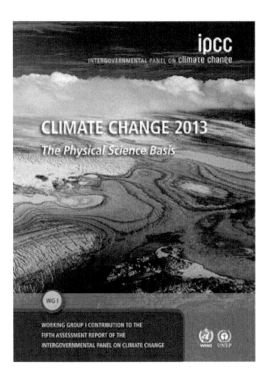

Most people, including most scientists, don't read IPCC reports, particularly those of Working Group I titled *The Physical Science Basis*. When they do, they either don't understand them or are shocked to find errors, omissions, deceptions and complete lack of data. The experience of German physicist and meteorologist Klaus-Eckart Puls in the statement at the start of this book tells you everything you need to know about the corrupted process of the IPCC.

The original IPCC was comprised of approximately 6000 people. It was gradually pared down to the current 2500, almost all of them bureaucrats, who work in four separate groups to prove the hypothesis that Anthropogenic Global Warming (AGW) is a fact. Working Group I (WGI) is comprised of 600 people who prepare the report titled *The Physical Science Basis*. The remaining 1900 make up Working Group II, which produces the *Impacts, Adaptation and Vulnerability Report*, and working Group III that produces the *Mitigation of Climate Change* report. They accept without question the conclusions of WGI. All their reports are withheld until a separate group of politicians and Senior UN Bureaucrats with a few hand-picked scientists produce the Summary for Policymakers (SPM). Their rules require publication and release of the SPM before the Science Report is released. The Summary must then go back to WG I to make sure their report confirms the Summary statements. It is like your boss writing a report for the company owner and then directing you to ensure that what is included happens. As IPCC Reviewer David Wojick said:

> Glaring omissions are only glaring to experts, so the "policymakers"—including the press and the public—who read the SPM will not realize they are being told only one side of a story. But the scientists who drafted the SPM know the truth, as revealed by the sometimes artful way they conceal it...What is systematically omitted from the SPM are precisely the uncertainties and positive counter evidence that might negate the human interference theory. Instead of assessing these objections,

the Summary confidently asserts just those findings that support its case. In short, this is advocacy, not assessment.

EVIDENCE OF THE POLITICAL OBJECTIVE

We've got to ride the global warming issue. Even if the theory of global warming is wrong, we will be doing the right thing...

Senator Tim Wirth, 1993

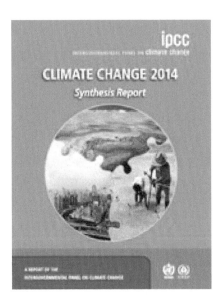

Wirth arranged for James Hansen of NASA to appear before a Senate Committee in 1988 at which the entire global warming scam was launched. They determined the warmest average day for Washington and scheduled the hearing for that day. The night before they went in and opened all the windows and turned off the air conditioning in order to make the hearing very hot and sticky as a simulation of what a warmer world would feel like. It was a disgusting orchestration and stage setting to begin the deception. His record showed Hansen was the person for the job.

JAMES HANSEN'S EXTREME MEASURES
(STILL COOKING THE BOOKS)

It is the type of politics that underscores the anger among voters and the low opinion they have of politicians. Wirth's "confession" is in a PBS Frontline documentary; he clearly thinks it was clever. Underscoring how the entire issue was only about politics.

Here are other expressed similar views:

> *One has to free oneself from the illusion that international climate policy is environmental policy...We redistribute de facto the world's wealth by climate policy.*

Ottar Edenhofer, Co-chair UN's Intergovernmental Panel on Climate Change working group on Mitigation of Climate Change from 2008 to 2015

> *This is the first time in the history of mankind that we are setting ourselves the task of intentionally, within a defined period of time, to change the economic development model that has been reigning for at least 150 years, since the Industrial Revolution.*

Christiana Figueres, Executive Secretary of UN's Framework Convention on Climate Change

No matter if the science of global warming is all phony... climate change provides the greatest opportunity to bring about justice and equality in the world.

C. Stewart, Minister of the Environment, Canada (1997-1999)

When we allow science to become political then we are lost. We will enter the Internet version of the Dark Ages, an era of stifling fears and wild prejudices, transmitted to people who don't know any better.

Michael Crichton, MD, Author

Our economic model is at war with the Earth... We cannot change the laws of nature. But we can change our economy. Climate change is our best chance to demand and build a better world... This [man-made climate change] is not about science, or health, at all.

Naomi Klein, Author, Social Activist

THE PRECAUTIONARY PRINCIPLE

Standard scientific method tries to disprove a hypothesis, but promoters of the AGW hypothesis tried to prove it. As Richard Lindzen said years ago about the global warming hypothesis, the consensus was reached before the research even began. Organizers of the deception knew that contradictory evidence would appear, so they built the precautionary principle into their directives.

As the cartoon about the dog indicates, the precautionary principle is the standard fallback position of environmentalists. It enshrines the idea that one must act anyway, just in case. The problem is that evidence is needed, otherwise more harm will come than good. Of course, another adage says to act in haste, repent at leisure. The following is Principle 15 of Agenda 21, the UN plan for the 21st century that enshrines this precautionary concept.

> *In order to protect the environment, the precautionary*
> *approach shall be widely applied by States according to*
> *their capabilities. Where there are threats of serious or*

irreversible damage, lack of full scientific certainty shall not be used as a reason for postponing cost-effective measures to prevent environmental degradation.

Notice it infers that the wealthier countries are to participate "according to their capabilities". It provides a way to avoid the need for scientific evidence because they get to define how much less than "full scientific certainty" is required. Compare this with Holdren's argument earlier for using the Constitution to justify abortions.

Indeed, it has been concluded that compulsory population-control laws, even including laws requiring compulsory abortion, could be sustained under the existing Constitution if the population crisis became sufficiently severe to endanger the society.

Who decides in either case? "They" do. It is a classic circular argument. They create the problem, determine when it needs attention, then create and implement the solution.

DISCUSSION

The global warming deception involved deliberately isolating human produced CO2 as the cause. The environmentally friendly reason used was to save the planet. The real objective was to eliminate capitalism and its industrial engine while reducing world population. The proposed replacement was socialism and alternative energies – the Green Agenda. It fails everywhere it is tried because socialism fails even when it is presented as a plan to save the planet. To paraphrase H.L. Mencken, *The urge to save the planet is almost always a false front for the urge to rule.*

If you want more detailed explanations,
please consider my book,
The Deliberate Corruption of Climate Science
available on Kindle or Amazon.

A Few Important Definitions

If you wish to converse with me, define your terms.

Voltaire

Weather: Atmospheric conditions at a place and time. It is the total of everything from cosmic radiation from space to heat from the bottom of the ocean and everything in between. A small portion is depicted in Figure 3.

Climate: Average of the weather over time or in a region. It is a statistic best summarized by Mark Twain that "climate is what you expect, weather is what you get."

Meteorology: Study of the physics of the atmosphere. This is considered essential training for weather forecasters. It is a subset of climatology.

Greenhouse Effect: The analogous claim that the Earth's atmosphere functions like a greenhouse. It allows shortwave sunlight in, which heats surfaces producing long wave energy, which is blocked from escaping by the glass. It is an inappropriate analogy.

Global Warming: Properly means that the trend of the global temperature is increasing. It is incorrectly and narrowly used to refer to the increase of global temperature due to human addition of CO2 to the atmosphere. In fact, it is only one of hundreds of changes that can cause a global temperature increase.

Climate Change: Something that has occurred throughout the Earth's history. What most people don't know is that the temperature changes dramatically and in relatively short time periods all the time.

Climatology: The study of climate in a region or change over time; a systems approach.

Climate Science: The study of one small piece of the vast and complex system of climate.

My book with elaboration and extensive references available from Amazon.com.

General References

Climate: Present, Past and Future

Hubert Lamb

This book, published in 1977, shows how much the IPCC restricted climate research and how much is not included in their work.

Meltdown : The Predictable Distortion of Global Warming by Scientists, Politicians, and the Media

Patrick J. Michaels

Global Warming in a Politically Correct Climate: How Truth Became Controversial

Mihkel M. Mathiesen

Man-Made Global Warming: Unravelling a Dogma

Hans Labohm et al.

Taken By Storm: The Troubled Science, Policy and Politics of Global Warming

Christopher Essex and Ross McKitrick

Global Warming: A Guide to the Science

Willie Soon et al.

The Satanic Gases
Patrick J. Michaels, Robert C. Balling

Global Warming and Other Eco-Myths
Ronald Bailey (Ed.).

The Greenhouse Delusion
Vincent Gray

Climate Change: A Natural Hazard
William Kininmonth

The Manic Sun
Nigel Calder

Global Warming: Myth or Reality?
Marcel Leroux

Unstoppable Global Warming: Every 1500 Years
Dennis Avery and Fred Singer

Blue Planet in Green Shackles
Vaclav Klaus

The Chilling Stars: A new Theory of Climate Change
Henrik Svensmark and Nigel Calder

The Hockey Stick Illusion: Climategate and the Corruption of Science.
A.W. Montford

Climatism: Science, Common Sense, and the 21st Century's Hottest Topic
Steve Goreham

Climategate: The Crutape Letters
Steven Mosher and Thomas Fuller

The Politically Incorrect Guide to Global Warming and Environmentalism
Christopher Horner

The Delinquent Teenager: IPCC Expose.
Donna Laframboise

Confessions of a Greenpace Droput
Patrick Moore

Green Gospel
Sheila Zilinsky

The Great Global warming Blunder
Roy Spencer

Heaven and Earth; Global warming, The Missing Science
Ian Plimer

Not by Fire but by Ice
Robert Felix

A Disgrace To the Profession
Mark Steyn

Climate Change: The Facts
John Abbot et al.

The Fable of a Stable Climate
Gerrit van der Lingen

The Moral Case for Fossil Fuels
Alex Epstein

The Neglected Sun, Why the Sun Precludes Climate Catastrophe
Dr Fritz Vahrenholt and Dr Sebastian Luning

CPSIA information can be obtained
at www.ICGtesting.com
Printed in the USA
LVOW01s1211010517

532690LV00015B/31/P